MATERIALS

DESIGN and TECHNOLOGY

AUSTRALIAN TECHNOLOGY STUDIES

MATERIALS
DESIGN
and TECHNOLOGY

RAY STEVENS

CAMBRIDGE
UNIVERSITY PRESS

ACKNOWLEDGEMENTS

Published by the Press Syndicate of the University of Cambridge
The Pitt Building, Trumpington Street, Cambridge CB2 IRP, UK
40 West 20th Street, New York, NY 10011-4211, USA
10 Stamford Road, Oakleigh, Melbourne, Victoria 3166, Australia

© Cambridge University Press 1992
First published 1992

Printed in Hong Kong by Colorcraft

National Library of Australia cataloguing in publication data:
Stevens, Ray (Raymond John)
Materials, design and technology.

ISBN 0 521 40412 6.

1. Technology. 2. Design, Industrial. I. Title.
(Series: Australian technology series).
600

A catalogue record for this book is available from the British Library.

ISBN 0 521 40412 6.

The author would like to thank the following people for their assistance and support throughout the compiling of this book:

My wife, Mary, and my daughters, Brooke and Jaye, for their invaluable understanding and support while I was so busy putting the book together; Mr Brian Ramsay, for reading rough drafts and providing many helpful hints and ideas; Mrs Grace Barbuto, for initial assistance with the typing of manuscripts; Margot Holden, editor, for her patience and continual assistance, and for her personal input into the 'Technology in Australia' and 'Research and Investigations' chapters; the students of Brunswick Technical School and St Bernard's College for their participation in the trialling of the worksheets and ideas used in the book, and especially those Brunswick students who gave me inspiration to put pen to paper in an attempt to assist students of Technology Studies; Mr Gary Underwood, for the opportunity to fulfil an ambition and for his patience and assistance when the going got tough during many changes to the original concept; Mr John Holley, for his support of the idea, his initial input and his final reading of the manuscript; and the many colleagues with whom I discussed ideas, issues and student needs.

Grateful acknowledgement is made to the following for their assistance in both providing suitable material and granting Cambridge University Press permission to reproduce it in this text:

BHP Minerals; Cambridge University Press, U.K. for drawings from *Design Briefs* by Lee & Aldridge (1987); General Motors Holden; Invertech Operations; Kambrook Distributing Pty Ltd; Mayne Nickless Limited; Philips Consumer Electronics; Qantas Airways Ltd; Rowville Secondary College; Royal Flying Doctor Service of Australia; Sanyo Australia Pty Ltd; Victorian Tourism Commission; Victa Lawn Mowers; the *Weekend Australian*; Telecom Australia.

Every effort has been made to trace and acknowledge copyright but in some cases this has not been possible. Cambridge University Press would welcome any information that would redress this situation.

Front cover photos: Scoopix Photo Library (skateboarder)
Rowville Secondary College (children at a computer)

CONTENTS

The AUSTRALIAN TECHNOLOGY STUDIES series has been written to assist all students participating in the study of technology.

This book, *Materials, Design and Technology*, contains an introduction and nine chapters related to technology education, and a case study in design — the Holden car story. Each chapter is self-contained and can be used as a separate learning unit but many aspects cross over and have links with material found elsewhere in the book. A glance through the comprehensive contents list will reveal some of these cross-overs.

Each chapter begins with a brief introduction and explains how the information will assist you in your studies. Revision questions at the end of each chapter will help you make sure you understand the information, and most chapters include student activities which will give you practical experiences of what you have learnt.

The book also contains several worksheets which may be of use to you in your classes. And at the end of the book is a Mini-dictionary containing many technical words and others which may be unfamiliar.

It is important for you to understand what is meant by the word 'technology'. Technology is a word which has numerous definitions but the one we will be using for the purpose of this book is as follows:

Technology is the use of a wide variety of skills, knowledge and creativity in attempting to find solutions for a particular need or problem.

I hope that throughout your time working in your technology classes this book will be of some benefit to

you. Remember, however, that a book is only as useful as the person reading it. If you rush through it without thinking, you will inevitably miss important aspects. This book will not only assist you with your technology classes, but will also give you an indication of the areas where technology studies have links with other subjects you study.

Good luck with your course. I hope you have many hours of enjoyment making your products, reading the book and participating in the activities.

Ray Stevens

vii

NOTES FOR THE TEACHER

The structure of this book is based on the belief that students can become self-reliant in developing their product ideas, in designing, drawing and evaluating them, and finding relevant information to assist them in this and to fulfil other work requirements of the classroom teacher.

WORKSHEETS

Throughout this book, I have referred to the use of worksheets. These are found as blackline masters at the back of the book and may be reproduced for use with AUSTRALIAN TECHNOLOGY STUDIES. They are intended as guides for students and it is not expected that you will *give* students the information to fill these sheets out. Students are required to research, investigate and discover — for themselves — the relevant information through the various methods discussed in this book.

The degree of difficulty in each part of the work outlined — and for each student — will vary. It will be determined by the type of brief you give students and the solutions that they decide on to overcome the problem or to meet the need given in the brief.

Much of the information you may require students to find out will be gained from your classroom discussions and practical demonstrations. It is then up to students to take note of these points and use them wherever and whenever relevant to their own needs. For example, in demonstrations of testing for tensile strength, one student may write down *all* the findings — if this directly affects a part of the project he or she is working on. Another student might merely note the *kinds* of tests that can be conducted, and record this for future reference.

EXCURSIONS

When taking students out on excursions, make sure they understand the *purpose* of the visit. Give them prepared question sheets which they can use to record information they may (and should) obtain. It is also a good idea to send — in advance — the question sheets to the company or business you are visiting. This gives the person who is showing you around an idea of your aims for the visit and he or she can then make sure the information is covered.

You could divide the group into three or four and provide each group with a different question sheet, including specific questions for them to answer. This type of guideline ensures that all students will gain a certain amount of information, will participate *actively* in the excursion, and share their findings when they return to the classroom. (When students make individual industrial visits, encourage them to be similarly prepared.)

INTRODUCTION: WHAT IS TECHNOLOGY?

People often lack an understanding of what is and what is not technology. Unfortunately, the word 'technology' is often only associated with things such as fast cars, computers, electronics, miracle medical equipment or the latest piece of CD equipment. In fact, these are just a small part of the world of technology that surrounds us. Technology is all of these things and much more.

Technology has been created out of the basic need of humans and is a major factor which sets humans apart from most other forms of life. For thousands of years it has helped in the supplying of our basic needs — food, shelter, transport, communication and clothing. This is still the case. However, technology has become broader and more complex in its use. Instead of merely helping to sustain life, it now prolongs it and makes it easier and more pleasurable.

It should also be noted that not all people believe that the ever-expanding world of technology is good for us. Many people lay the problems of today's world squarely at the feet of the rapid technological developments which have taken place during more recent times. Unemployment, the destruction of the environment, even street violence and many other issues, are blamed on the over-development of technology. An important issue may well be the need to educate the makers and users of the products of technology to make sure that the development is controlled for the betterment of all concerned.

The manner in which people use technology during the next decade will determine the type of world that following generations will inherit. Responsibility lies not only in the hands of you and your teachers but also with large companies, businesses and governments to produce sound guidelines for all to work within. It is hoped by the time you complete your technology education classes you will have a broad understanding of the world of technology and that through your learning experiences you will be in a position to make sound judgements about the use of technology in today's world.

1

TECHNOLOGY IN AUSTRALIA

Although Australia is a relatively young country in the field of modern technology, it can hold its head high when compared with others in the development of many useful products used world-wide, as well as a wide variety of products for the local market.

If you are to understand technology in Australia you need to understand who first used technology, why it was used, what technology was developed over the years and how it fits into Australia's economy today.

This chapter will look at the development of technology in Australia before and since white settlement — including an outline of the importance of agriculture, mining and manufacturing in this development. It will look at the remarkable contribution of our inventors and innovators, and, finally, examine the role of technology in the future of Australia.

The focus of the chapter will be:

- to compare the use and development of technology by Aborigines and European settlers
- to develop an understanding of technology and its use in Australia during the past 200 years
- the role (and products) of Australian inventors and innovators
- Australia's current position in the world's manufacturing market-places
- the future of technology in the modern world.

ABORIGINAL TECHNOLOGY

The first inhabitants of Australia were the Aborigines who travelled from Asia at least 40 000 years ago. They developed their own form of technology to assist them in their way of life.

Fig. 1.1   Aborigines fished and hunted using canoes, boomerangs and spears made from natural materials.

Where and when food supplies were plentiful they remained in one place and built substantial dwellings. At other times they moved within their familiar tribal boundaries, following the seasons and gathering and hunting food where they knew it could be found. It was therefore necessary for them to travel lightly and use the available resources to their maximum. This is most notable in the type of housing they lived in and the fact that their tools were often used for more than one particular purpose.

Fig. 1.2   Aboriginal shelter (the humpy). Note clothing made from possum skins.

During the thousands of years they inhabited Australia, Aborigines learned how to use the elements and natural resources. They constructed traps to catch fish from the sea, creeks and rivers. The woomera (spear thrower), boomerangs, and spears made from tree saplings and stone, were used to hunt kangaroos, wallabies, lizards and other animals and birds. Digging sticks made from wood were used to dig up yams and various plant roots. Carrying bags for the collecting of fruits and berries were also made from the natural resources that surrounded the Aboriginal campsite. Fires were used to clear hunting grounds and encourage the growth of new plants (to attract more animals), to clear campsites, cook food, and harden wooden spear heads.

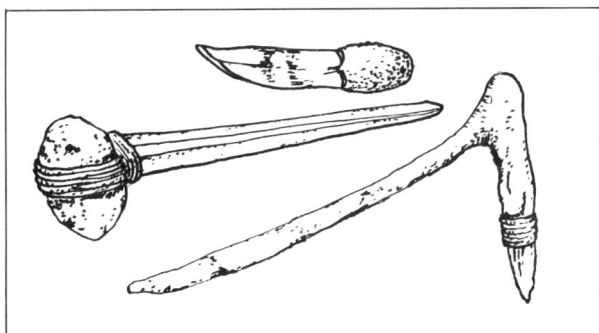

Fig. 1.3   Tools made for hunting and building shelters

The biggest asset which enabled Aborigines to survive in even the harshest wilderness areas was their understanding and knowledge of the local environment and resources. They understood the seasonal changes (the frequent occurrences of droughts and floods, etc.), the habits of the native animals and plants, and other environmental factors. This knowledge was learnt through practice and acute observation and was passed down through the ages in song, dance, painting and rituals.

EUROPEAN TECHNOLOGY

In 1770 Captain James Cook charted the east coast of Australia and claimed the land for Britain. Eighteen years later, the First Fleet landed and established the first white colony. This was the first step towards the development of Australia as we know it today.

The First Fleet's arrival brought with it many forms of technology which had been developed over thousands of years throughout the European world. Sailing ships, carriages, tools and equipment for constructing roads and housing and, of course, guns to maintain law and order were just a few of the technologies which were introduced at this time.

In the early days, the British settlers did not understand their new country and its natural environment — so very different from their homeland — and were unwilling to learn anything from the original inhabitants. (As we now know, this lack of knowledge eventually exacted a heavy price on the land.) The settlers imported from Britain all the technology they required to satisfy their needs, and this did not change for many years. However, it was found that many forms of equipment were unsuitable for Australian conditions, so the settlers were required to develop their own forms of technology or to import technologies which had been developed in other countries. An example of this is the introduction of Cobb & Co coaches which were imported from America in the early 1850s to replace the inadequate English-style coaches which had been used in regular service from around 1814. The American-built coaches were far more suited to Australian conditions since they had been developed for the rugged American West.

Settlers eventually learned to take into consideration the unique natural conditions of Australia and combine this with the knowledge and skills of the European technology with which they were familiar. The end result of this development was the beginning of our own Australian technology.

SETTLEMENT EXPANDS

The five main areas of technological development from this time were based around the need to feed, house and clothe the people of the colony, to transport goods and equipment to the outlying areas which were growing rapidly, and to communicate between these outlying areas and the main settlements.

White settlement expanded at a great rate between the 1820s and 1850s. During this period free settlers were migrating to the 'new' country in droves. They grew crops, bred sheep and ran cattle farms. Many later ventured into the interior of Australia and beyond to set up new settlements or stations in remote areas.

The discovery of gold in the middle of the 19th century provided further impetus to the development of Australian technology. Clerks, shepherds, farmers, sailors and professional people all packed in their jobs in the hope of discovering gold and being set up for life. It could also be said that it was the gold rush days which set the scene for much of the characterisation of the Australian way of life. The term 'digger' can be traced to the gold fields along with the 'mateship' which developed out of the need for people to rely on each other in times of hardship. The organisation of the Eureka Stockade is one example of this.

The settling of the land and the discovery of gold brought with them the added need for skilled people to work in and further develop the industries required to supply the settlements or gold fields with equipment, shelter, food, transport, clothes and the ability to communicate with each other.

MINING

The mining industry has always been one of our major industries — from the time of the first discovery of coal at the mouth of the Hunter river in the late 1700s, through the gold rush times in Victoria and Western Australia, to the establishment of large mining companies, such as BHP based in Broken Hill.

Fig. 1.4 In the early days mining was very labour-intensive work.

The mining of coal for fuel, iron ore and copper for use in manufacturing industries, and gold for personal gain, all pushed forward the rapid development of technology in the last century and still plays a major role in this area and within Australia's economy today.

Fig. 1.5 Today, large machinery has taken over much of the work in the mining industry (BHP Minerals Division)

AGRICULTURE

The farming or agricultural industry is diverse and widely spread throughout the length and breadth of Australia. The many forms of farming ranging from growing crops such as wheat and sugar cane, to cattle farming supplying beef and dairy products, sheep farming for meat and wool, and the numerous fruit-growing orchards across the country have played no small part in the development of Australia and its technology during the past two hundred years. The exporting of our primary produce, especially wool, has been very important to our economic growth. In the best times the country was said to be 'riding on the sheep's back'.

MANUFACTURING

By the end of the 1930s, Australia's manufacturing industries were well established. From the Second World War to the '70s they generally flourished. With a growing population (from immigration and the post-war 'baby boom') came new skills, an increased demand for goods and services (schools, hospitals, etc.) and the expectation of a higher standard of living for all.

The 1950s and '60s were the boom years and Australia was among the leading nations in the development and improvement of (some) machines, equipment and manufacturing processes. But as a society we did not encourage our inventors and innovators, and in the '50s about one third of our science graduates went overseas to find work. Most of the money from our exports still came from the sale of primary produce (particularly wool, wheat, etc.) and raw materials (iron ore, coal, etc.).

At the same time — while we were 'riding on the sheep's back' economically — our manufacturing industries were protected against competition from similar but cheaper goods from overseas by tariffs. These are government charges on imported goods to make them more expensive so that equivalent Australian products can compete.

Tariff protection and government restrictions on imports over many decades also meant that many of our industries did not have to be very efficient to survive. They could continue to use old techniques and outdated equipment, pay reasonable wages and still stay in business. Meanwhile, European and Asian competitors in the market-place built up new industries using cheaper labour, new machinery and investment capital (money), particularly from America. Eventually, over recent years, tariffs were reduced and the competition caught up with us.

In the 1990s we can no longer rely on our traditional products to bring in the money we need to maintain our present high standard of living, or on tariff 'walls' to protect us from competitors. We have to restructure and re-equip our existing industries to make them more efficient and we will have to develop new ideas, new skills, new products and new markets (buyers) for those products.

Governments, industry leaders, economists, educational establishments and other people are all trying to work out ways of encouraging this development and attracting investment in Australian manufacturing. There is much talk of the need to develop more 'value-added' industries — that is, industries which add further value to our abundant raw materials *before* we sell them. BHP already adds value to some of our iron ore by processing it and selling it as steel; products made from that steel (e.g. Colorbond) become *more* valuable — and earn more money when sold overseas.

All-out bid to boost exports

By PAUL DOWNIE

THE potential for boosting the nation's exports – through better returns by replacing raw-material exports with products made with them, and overcoming restrictive agreements – will be examined in a dual inquiry by the Industry Commission.

The two inquiries, announced yesterday, seek public submissions on how to boost Australia's export competitiveness and identify barriers to trade in overseas countries...

The announcement of the inquiries followed the release yesterday of Bureau of Statistics figures showing that

Forge new trade links

Japan continues to be Australia's strongest trading partner in merchandise goods.

The March 1991 quarter figures of Australian merchandise exports and imports with major trading partners show that exports exceeded merchandise imports by $673 million for the quarter.

It is the fourth successive recording of a trading surplus following seven quarterly deficits...

The most improved trading relationship occurred with Korea, resulting in a trading surplus to Australia of nearly $800 million...

While Japan continues to be Australia's main destination for exports, the nation has showed a clear preference for importing goods from the United States which provided, during the March quarter, 24 per cent of all its imports.

Australia was a clear loser in terms of merchandise trade with the US, showing a $1.73 billion deficit between the value of US imports and Australian exports to the US...

The Industry Commission inquiries are aimed at helping Australian business extend and forge new trading links with overseas partners.

The inquiry into the pricing of locally produced raw materials, intended to cultivate awareness of the potential for value-adding enterprises, is to examine:

WHAT raw material prices limit further processing in Australia in favour of processing abroad.

WHY raw materials are sold on to the export market at prices below those on the domestic market.

All materials, except wool, will be covered by the inquiry. The IC will attempt to identify whether prices are an impediment to adding value to raw materials in Australia and the export of manufactured goods...

Source: *The Weekend Australian*, 29 June 1991 (p.3)

Fig. 1.6

AUSTRALIAN INVENTION AND INNOVATION

The world of the inventor and innovator is not an easy one. Many hours of hard work and exhaustive investigation is put into attempts to develop products and ideas which can be used to benefit the world around us. The development of technology has continued throughout the years because of inventors, designers and creative people displaying imagination and innovativeness in producing new and exciting products. As long as there are products or needs that represent a challenge, these people will strive to improve those products and people's lives.

In Australia's modern history there have been many changes in the way our country has developed, due in no small manner to the unique nature of the environment. As stated earlier, much imported technology was of little use and new solutions had to be found for local problems. In the beginning this was difficult; however, as people became accustomed to the conditions, they began devising equipment suitable for working the harsh land. The stump jump plough is a good example of this.

Inventions such as the stripper/harvester, which was used to harvest grain five times faster than before, and the winged keel which helped Australia's yacht to win the America's Cup in 1983, although spaced many years apart, helped to put Australian innovation in the forefront of international technology. Many other innovations and inventions such as the Flying Doctor Service and the rotary clothes line were produced to assist with unique Australian needs.

Fig. 1.7   The winged keel which helped Australia win the America's Cup
(Qantas)

In Australia we have often been slow to show interest in, recognise or support financially the inventions and innovations put forward by our own inventors, industrial designers and scientists. In recent years (as in the past), inventive Australians have produced plenty of good ideas for products, particularly involving so-called 'high-tech' techniques (computerised technology) but have found it difficult to persuade other people to invest their money to make the idea a reality. Newspapers often carry stories of innovative ideas that have been 'sold' to more adventurous overseas companies because the creator of the idea cannot get local backing to produce it in Australia.

It takes imagination and courage, too, to take the kinds of risks that are necessary to research, develop, test and make the product, and then to market (sell) it overseas. The government and broader community need to be supportive with policies and the purchasing of these products in an effort to assist in taking our flagging manufacturing industries through the '90s and beyond with new hope and enthusiasm.

WHERE TO NOW?

All the innovative products listed on page 7 have been invented, produced or developed in Australia. These products show what can be achieved if the support and incentives are high enough to encourage our inventors and designers to produce new and innovative products.

Technology right throughout the world has hit boom times. The microchip, compact disc, video recorder and the bionic ear are just a few products which are far advanced compared with their forerunners of only a decade or so ago. The speed at which new products are developed and appear on the market is sometimes frightening — not only to the lay person who has little understanding of the technology but also to people working in industries which are being dramatically affected by the indirect consequences of the new 'technological age'.

Many new machines are capable of handling a lot of jobs which were initially done by people and they can carry out those tasks to a higher standard, in a shorter time and more efficiently. This has resulted in many people being made redundant. On the positive side, these machines have assisted in reducing many work-related injuries as well as replacing humans altogether in many workplaces which are conducive to health risks. Robots (used extensively in the motor car industry), word processing machines, ergonomically designed furniture and computerised machines and equipment all assist us in the carrying out of our daily work with much more ease than was previously possible.

Fig. 1.8   Transport developed to cover great distances
(Mayne Nickless Limited)

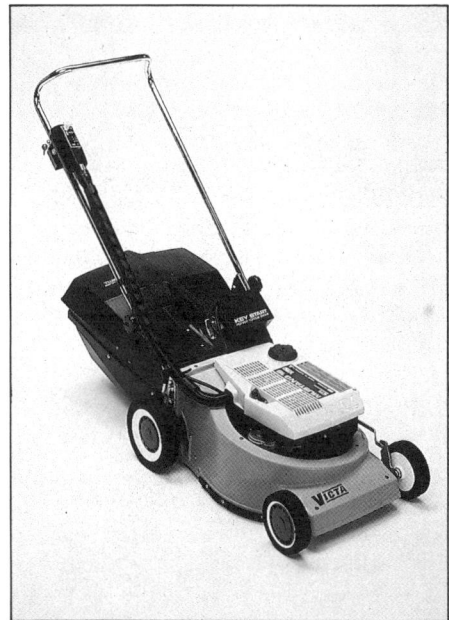

Fig. 1.10   The Victa mower

Fig. 1.9   Aircraft used for the Royal Flying Doctor Service

Fig. 1.11   Many forms of technology improve our daily life. But do they have only good effects? (Philips Consumer Electronics) (Sanyo Australia Pty Ltd)

All of this technology has been at a cost and it has been only in recent times that our society has begun to come to grips with the many problems which have been caused through the misuse of technology or the lack of forward thinking in the development of new products. Aerosol cans, leaded petrol and non-degradable plastics are examples of things which, when first produced, were believed to be of great value in making our lives easier. We all know the problems we live with today because of the development of these products as well many others. Our awareness is so much higher, and as a society we are beginning to question the value of some of the products being introduced to our world. The saving of the forests, the cleaning up of our atmosphere and waters, the conserving of our non-renewable resources, and the finding of alternate technologies lie in the hands of us all. The manner in which we control technology will determine the type of world we live in and future generations will inherit.

REVISION QUESTIONS

1   What does the word 'technology' mean?

2   Circle either TRUE or FALSE:

   a   The Woomera is an Aboriginal musical instrument.   T   F

   b   Technology is concerned only with computers, robots and technical machinery.   T   F

   c   Most of the tools used by the first settlers were Australian-made.   T   F

   d   Technology is only harmful when used incorrectly.   T   F

   e   The boomerang is a piece of technology.   T   F

   f   A person who makes new products is called a scientist.   T   F

   g   The first stripper/harvester was invented in Germany.   T   F

3   Name the *two* main industries that assisted Australia's development in the early days of European settlement.

4   Why was it necessary for the white settlers to develop their own technology?

5   Name *four* areas of human need which are continually being improved through technology.

6   Explain the difference between an invention and innovation.

7   Explain how tariffs protect local industries.

8   Why is it that some good Australian products are developed and made overseas?

STUDENT ACTIVITIES

Select *two or more* of the following five activities. They relate to the early part of this chapter.

1   You have read about technological advancement in Australia over the past 200 hundred years. On an A4 sheet, present a flow chart which represents the information and presents the ideas.

2   Produce a simple flow chart which shows the development of one of the products listed below. Indicate any major developments in the materials used and the design of the product. (Use magazine cuttings if you are not confident in your drawing skills.)

- the telephone
- home building
- a piece of farming equipment
- cooking ovens
- television & video
- an item of sports wear
- bicycles
- raincoats
- aeroplanes
- radios and stereo units
- Housing Commission homes
- the Holden car
- one form of public transport
- roads
- surf boards
- writing pens
- calculators
- swimwear
- school uniforms
- the automobile
- cooking utensils
- garden tools

3   Write a summary which explains and describes the effect that technology has had on the overall development of Australia during the past 200 years.

Include the good and bad effects of technology throughout this period, and say where you see technology taking us during the next 20 years.

4   In each of the following groups, circle the *odd one out*. (You may need to use the library for this one.)

   a)   Ben Lexen   John Ridley   Sir Robert Menzies
   b)   Dowell Industries   Hills Industries   Ford
   c)   Kambrook   General Electric   Gold Star
   d)   Barry Jones   L. Bandt   J.M. Holden

5   Australians have been involved in designing, inventing and producing new and innovative products for a long time. Investigate *one* product which has been designed or manufactured in Australia. Write a conclusion on the product's ability to carry out its intended function, and your opinion of the design, manufacture and appeal of the product.

   Use the PRODUCT RESEARCH worksheet (provided on page 70) to help you. The list below may help you in your selection. Information relating to these products and innovations can be found in: *Australia's Best*, Australian Design Council and the Advance Australia Foundation, 1988, and *Technology in Australia 1788-1988*, Australian Academy of Technological Sciences and Engineering, Melbourne, 1988.

- Coolgardie safe
- stripper/harvester
- Dolphin Lamp
- the woomera
- Hills Hoist
- Sunbeam Monitor iron
- Furphy's Water Cart
- Holden car
- steel-framed housing
- Cochlear Bionic Ear
- T.21 chair
- Kevron Key Tag
- Nomad aircraft
- stump jump plough
- Kambrook Power Board
- Victa mower
- aluminium windows
- cordless electronic kettle
- Sydney Harbour Bridge
- Pulsar car battery
- Stackhat
- Oates mop bucket
- Camatic Clerical Chairs
- Road Train

PRODUCT RESEARCH

PRODUCT ......*ELECTRICAL POWERBOARD*...

Company or person responsible for development: ......*Kambrook*.....

....

When was the product developed? ....

Why was the product developed? *So people could plug in and use more than one item at a time. Some powerboards have six (6) points to plug into.*

Was it developed especially for Australian conditions? If so, explain. *Not specifically. Although developed in Australia*

How was the product developed? *The person who established Kambrook was an electrician, he was sick of plugging and unplugging power tools, he worked on ideas from the double adaptor.*

Who used (or uses) the product? *We use them at school for the computers and printers. They are also used in garages and homes.*

What are the product's strengths? *Assists people who require a number of items to be used at the same time. Relatively cheap*

What are the product's weaknesses? *Some I have seen are bulky and have large plugs.*

CONCLUSION *I believe this product is well worth buying. They can get in your way and you need to watch where you place them or you can trip over the leads. It is made from plastic. They can also be purchased with power surge guard or circuit breakers.*

Fig. 1.12   A completed PRODUCT RESEARCH sheet

CHAPTER 2

STUDENT ORGANISATION

The type of work you will be involved in during your technology education classes will be a mixture of theory and practical applications. It is important that you are well organised and aware of the tasks which are required to be completed for assessment purposes. Throughout this chapter you will be given many hints on these matters. Look through them and select which ones you feel will be most useful for your particular work units.

This chapter will focus on:

- things you need to be aware of before commencing your technology education classes
- what tasks you will be required to complete during and on completion of particular work units
- some ideas on organisation and planning which may assist you
- how to organise a folio containing related worksheets and notes
- the use of journal/logbook writings.

WORK REQUIREMENTS

Before starting your course, it is essential for you to understand what work will be required of you in learning about technology and for assessment purposes.

The requirements may be defined by the teacher or by an outside body such as a State education department. In some cases the teacher may discuss with you the requirements that are to be met and you may have input into decisions about such aspects as the size of the project, the type of material to be used, how a product is to be finished, and so on.

Many of the tasks you will be required to carry out will have deadlines for completion. These are sometimes called *due dates*. Tasks must be completed by a given time and be of an acceptable standard to your teacher.

It is important that you make sure you are informed about the expectations and organisation of each unit of work you undertake. It would be advisable for you to write down all of the work requirements for each unit of work, including the due dates for each task, and to keep a record of these for reference at a later date.

Some questions you may wish to ask are:

- What are the requirements of the unit?
- How is assessment related to the work requirements?
- What does 'satisfactory completion' mean?
- What is required for folio work and note taking?
- When is the work to be handed in?

- What standard or quality of work is required (for example, layout of drawings, format of presentations, neatness, etc.)?
- Do I have to conform to any school or departmental policies (for example, 75% attendance required in class)?
- Which tasks are to be assessed?
- How will the tasks be assessed?

ORGANISING YOUR FOLIO

In your technology classes you will need to be very well organised. Not only are you involved in producing practical work but you need to maintain a folio — a file or store of drawings, research materials, related theory notes, photographs, surveys, letters, homework sheets and research/investigation assignments. With this in mind, consideration must be given to the type of folder you are going to use for this purpose.

The type of folder which contains plastic envelopes is generally the best as you are able to keep your notes tidy and clean as well as move the papers in and out readily (see Fig. 2.2). This type also allows for easier collection by your teacher. You need to hand in only the immediate work being corrected and your teacher is able to place it straight in a folder which is similar to the one you are using, thus enabling the work to be protected at all times.

You should *date* and *file* each piece of work as it is completed. If you need to use a piece of finished work, such as a drawing, it would be advisable to photocopy it.

This will prevent the original becoming damaged, possibly affecting your assessment.

Another idea is to divide your folio up into *sections*. This will be helpful when looking for information in a hurry and it will also assist your teacher when checking and correcting your folio. The various sections could be named as follows:

- Ideas and sketches
- Drawings and plans
- Design notes
- Design analysis
- Materials information
- Materials testing
- Production plans and notes
- Research and investigation
- Reports
- Project evaluations
- Homework
- Journal/logbook

ORGANISING YOURSELF

A good way to ensure that your folio is kept in order and up to date is to set yourself a plan for *when* and *how* you are going to file your information. Write it out and keep in your folder. The plan suggested on the next page may give you some ideas to work from.

Fig. 2.1   Always file handouts and research information immediately.

Fig. 2.2

BEFORE YOU START

- Obtain a suitable folder for storing your work.
- Check on the due dates for each task.
- Check on assessment and reporting methods.
- Make sure you recieve all relevant worksheets.

EVERY SESSION

- Make sure you have the correct equipment required to complete your work.
- Fill in your journal/logbook after each session.
- Plan your activities for the next session. Record this in your planning sheets or journal.
- Place all handouts, research notes and worksheets in the appropriate section of your folio.
- Replace notes and drawing sheets used during the session.

EVERY WEEK

- At the beginning of the week, check with your teacher on how you are progressing with your work.
- Keep records of any relevant information you gathered about your project.
- Make sure all work from the previous week is up to date.
- In your journal/logbook, note any alterations to your project.
- Complete all homework tasks.
- Plan your practical activities for the coming week and note any tools and equipment you will require.

KEEPING YOUR JOURNAL/LOGBOOK

A journal or logbook is used to record your progress throughout each unit of work. It will assist you in being well organised and prepared, and will save you time otherwise wasted on trying to remember where you are up to each time you enter the classroom.

If it becomes necessary to alter a part of your project, use the journal to show any alterations and explain the reasons behind having to make them. Each time you make an alteration, get your teacher to check it and approve your action by signing the sheet.

Your journal should also record the amount of work you do each week and your plans for the next week's work. Include notes on any new tools, equipment or processes you have learnt about.

Entries should be on a regular basis; how often, depends on you and your teacher.

Figure 2.3 shows one way to keep records. A blank JOURNAL ENTRY sheet is provided on page 71.

REVISION QUESTIONS

1   Why is it considered important to gain an early understanding of the work requirements for your course?

2   What is the main use of the folio?

3   Circle TRUE or FALSE:

   a   The journal is used to record your progress throughout a unit of work.   T   F

   b   Planning ahead will assist you when working on your projects.   T   F

   c   It is all right to alter your plans without consulting your teacher.   T   F

4   The journal is used to help you record information. Name *three* types of information which may be recorded in your journal.

5   Name *four* types of information or materials which could be stored in your folio.

JOURNAL/LOGBOOK ENTRY

NAME .... *Petra Loyal (11B)* .... DATE *16-3-92*

PROJECT .... *Tracksuit* ....

Work done this week/session:

I have been very busy cutting my material from the patterns I made up last week. I needed to check all of the parts and make sure they were correct. I also enquired about how I am going to join the parts and Mrs Haily is going to demonstrate this next lesson.

Alterations:

The only alteration I have made is a change of colour (to the school colours, I did this last week.) I made this change as I have been picked in the school athletics team and would like to use the tracksuit when I'm competing for the school.

Plans for next week/session:

Mrs Haily is going to demonstrate the way to sew the parts together. I have booked a machine for next lesson and hope to have part of the tracksuit put together then. I also need to purchase two zips for the legs as the school does not supply these.

Teacher's signature: .... *Mrs K. Haily* ....

Fig. 2.3 A completed JOURNAL/LOGBOOK sheet

CHAPTER 3

PROBLEM SOLVING IN DESIGN

A product is usually designed and made to satisfy a need or to solve a problem. In this chapter you will learn about — and practise working through — the *problem-solving process* in design. This process involves a series of basic steps.

You will be introduced to *design briefs* and how they assist people in producing products, and you will learn how to construct your own.

The following chapters give more details of each stage of the problem-solving process. This chapter is intended as an introduction to the *process* only.

The focus of this chapter is:

- to introduce you to the process for finding solutions to a given problem or need
- to help you to understand the steps to be followed when working through the problem-solving process
- to show the relationship between problem-solving and technology studies
- to introduce you to various types of design briefs.

The local council has a problem. The swings in some neighbourhood parks have been vandalised and need to be replaced. None of the available models satisfies the council's need for a sturdy, safe and vandal-proof swing.

They approach a suitable manufacturer and 'brief' him on the problem — that is, they give him an outline of the problem and their needs. He hands over the problem to a designer, providing her with a written *design brief*. This specifies the criteria (the specific needs) which must be met to satisfy the client, and any constraints or limitations on the project, for example, costs. In its simplest form, the brief might look like this:

DESIGN BRIEF

CLIENT:
Nurglewood City Council

PROBLEM:
The council needs to prevent damage to play equipment in local parks. Design a swing according to the following criteria:

CONSTRAINTS:
- The swing must be vandal-proof.
- It must be made of plastic.
- It must be designed for use by children aged 5-10 years.
- Each swing must cost no more than $500.00 (retail).
- It must pass all recognised safety tests.

Now the designer has to come up with some ideas, develop these ideas (perhaps making some models to show the client), and test the ideas until a final model is decided on and the swing is eventually made.

If all the criteria and specifications have not been met (for example, the swing fails the safety tests), the client will not be satisfied and it's back to the drawing board! Of course, good designers try to solve problems *before* the product is made.

THE PROBLEM-SOLVING PROCESS

The problem-solving process involves carefully working through a number of steps. While the steps can be listed in a particular order or sequence (see below), it might be necessary to move back and forth throughout the forming of the solution. This may occur because of some unforeseen problem concerning materials or processes, and alternatives may have to be found.

Problem-solving in design also involves the use of many techniques: drawing, planning, researching information, testing materials and processes, product analysis, quality control and evaluation. These are used to assist in the overall development of the project.

Descriptions of the problem-solving process vary depending on where and why it is being used. However, the five steps or stages listed below are common to most. They will help you to work through your own design problems.

1   Identify the problem, task or need.

2   Formulate the brief.

3   Develop, test and refine ideas.

4   Produce the product (manufacture, construction, fabrication).

5   Evaluate the product.

IDENTIFY THE PROBLEM

The identification of the problem or specific need is the first important step along the pathway to achieving a solution. Once the problem or need is identified, a short statement is written to explain the problem, or a rough sketch is drawn to show the problem clearly.

Fig. 3.2

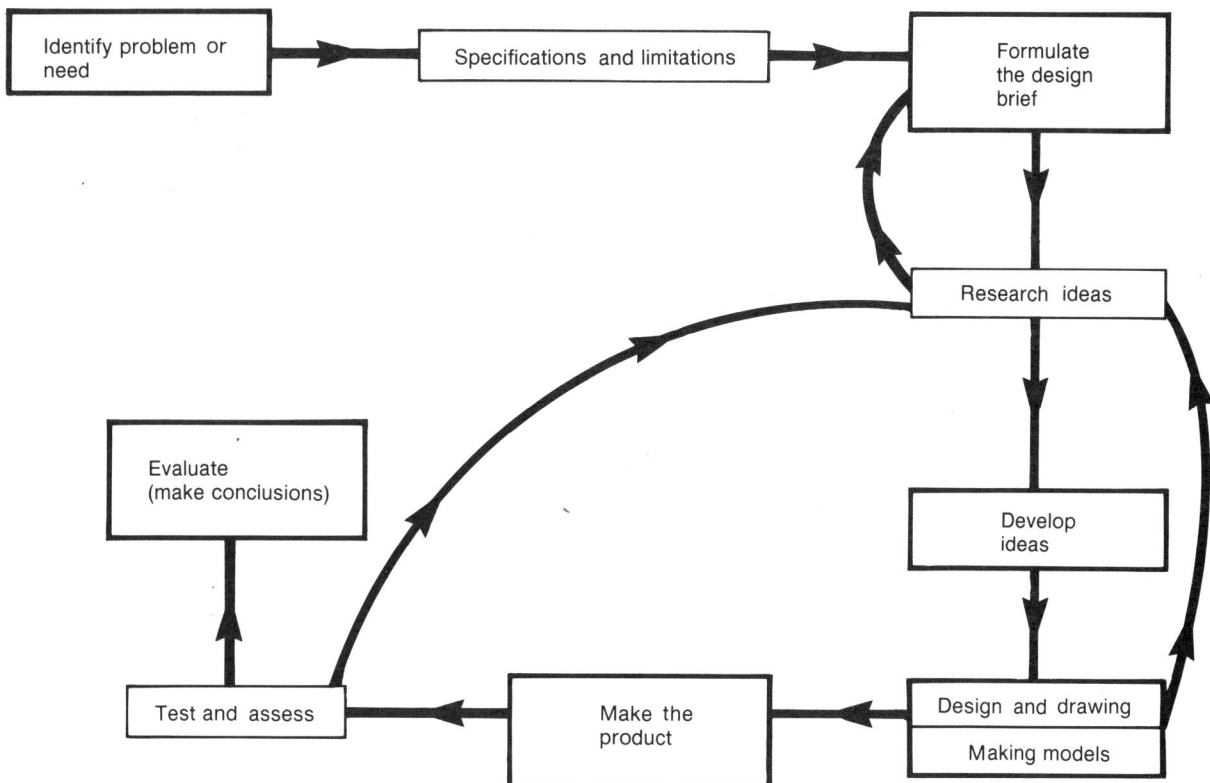

Fig. 3.1   The problem-solving process

FORMULATE THE BRIEF

The problem or need is easier to tackle if the design brief is clear and precise. If the brief doesn't give the correct specifications and requirements, the person designing or making the product will be unable to complete the work to the required standard.

The brief must state the problem to be solved, the purpose of the item, and its exact requirements. The statement should take into account considerations such as where the item is to be used, who will use it, any constraints or limitations on the project (e.g. costs, time limits) and the availability of materials, equipment and other necessary aids. (The design brief and its use in the design process is discussed further below.)

DEVELOP, TEST AND REFINE IDEAS

After identifying the problem and gaining a clear picture of the needs and requirements from the design brief, it is necessary to gather as much relevant information as possible to develop ideas for the project.

Information comes through research and investigation. This is achieved in various ways: by asking questions and discussing the problem with people who may be able to help; reading technical journals and books; writing to companies and business firms for assistance; looking at similar products; making models; using the school or local library. And, of course, you will try to come up with ideas of your own. (You may see something on TV and feel you could improve on it.) Remember, every piece of information may be important so don't discard anything unless you are totally sure it will be of no use.

Once you have gathered the relevant information, you will need to draw up your ideas. Always sketch two or three possibilities so you will have a broader base to work from if one of the ideas turns out to be unsuitable.

Analyse your ideas and select the one you believe will be best in assisting you to solve the problem at hand. Make sure it will cover all of the criteria (the specific requirements outlined in the design brief). If possible, make a model or prototype of the product to check whether it functions correctly.

You may also need to test the actual materials and processes that are to be used, to ensure there will be no problems with the making of the product. If you strike trouble you will need to refine your ideas and work back through the process until a solution is found.

PRODUCE THE PRODUCT

In your research and investigation work you will have found out about specific methods which will be most suitable for your project. Use the relevant information to assist you in putting the project together.

No matter how good the design is, if you don't use the correct tools and equipment, follow correct work procedures, and work accurately and neatly, the project is destined to fail. Pay particular attention to teacher demonstrations and instructions as they will help you learn the skills and techniques required in making your products to a high standard.

Fig. 3.4

EVALUATE THE PRODUCT

Evaluation should take place throughout the design and production process. This means making judgements on how the project is going and whether alterations are required. You will often need to rethink your initial ideas and plans because of unforeseen problems along the way. The alterations may be only minor and cause no real hold up, or they could be more serious and require a whole new approach to the project. By evaluating your work on a regular basis, you will hopefully pick up any major problems before you have progressed too far. Remember, it is always handy to have more than one idea or plan at your disposal in case problems do eventuate.

In your *final* evaluation the main question you need to answer is whether or not the product carries out all of the functions it was initially intended to do.

Fig. 3.3

Fig. 3.5

DESIGN BRIEFS

Design briefs can vary in their purpose and application. Some briefs are very basic while others contain a large amount of complex information. In your initial technology work, your teacher may give you a brief which contains a considerable amount of information and instructions. The brief might give details about the materials you are required to use and how the product is to be joined; it might specify *constraints* (limitations) on the size; it might include the actual drawing to work from. As you become more experienced with using design briefs and you acquire more skills in your technology classes, you may be given the chance to formulate your own briefs.

The three types of briefs you will most commonly come in contact with during your course will be initiated by your teacher, yourself or a third party (a client).

TEACHER-INITIATED BRIEFS

Your teacher may give you design briefs that have already been prepared. These are generally used to give you experiences in particular aspects of technology that you are required to learn — mechanisms, structures, tools, equipment, materials, various production processes, design — and, of course, practice in problem solving.

STUDENT-INITIATED BRIEFS

This is a brief that you write yourself to accommodate a personal need or to solve a personal problem. You are involved in the total process — from formulating the brief, to designing and costing the project, through to the making and the final evaluation of the product.

Fig. 3.6 Formulating a design brief

Fig. 3.7 Student-initiated design briefs

CLIENT-INITIATED BRIEFS

You may be required to find a solution for a need or problem for someone else, for example, another teacher at your school, a member of your family or a friend. The client may put in a request for a particular need that he or she wishes to be fulfilled. You would then be required to gather the necessary information from the client, formulate the brief, develop ideas and make the product. You would also need to consult with the client regularly throughout this process.

CONSTRAINTS AND LIMITATIONS

Whenever a design brief is put together there will usually be certain elements which will be beyond your control. These will place constraints on you and limit what you can do in your project. The following list represents the constraints which you are most likely to have to take into consideration in your own work.

TIME – The solution may need to be found before a certain date. That is, you will have to work to a deadline.

PROCESSES – The equipment available may limit your choice of methods for making or finishing the project.

MATERIALS – The choice of materials may be specifically defined in the brief, or they may be limited to those available in the classroom. None of the available materials may be suitable for making your particular item and you may have to rethink the whole project.

COST – You may be given a budget or cost range to work within. (Remember the Nurglewood Council's swing?)

REVISION QUESTIONS

1   What are the *five* main steps used in the process for solving problems?

2   Explain what a design brief is used for.

3   What is a teacher-initiated brief?

4   Name *two* sources of information which will help you in thinking up ideas for products.

5   Why it is important to test and check your ideas?

6   Why is evaluation of the product essential?

STUDENT ACTIVITY

The following activity will assist you to understand the problem-solving process.

How often are you hounded about keeping your desk or dressing table tidy? We always seem to have pens, pencils, erasers, pins and all sorts of bits and pieces all over the place on our desks and tables. Just to keep the adults off our backs, let's see if we can come up with a viable solution to this problem by designing a container to keep the bits and pieces in.

Read again the problem-solving process outline and use what you have learnt. Write a *design brief* (see page 15), then *design and develop a mock-up* of a container that can be used to keep your desk or dressing table tidy. (The final product would be made of wood, metal or plastic.) Read all the following information before you begin.

When creating your design brief, make sure you include at least *four* of the items listed below:

- 12 coloured pens (packet)
- 2 video tapes
- 2 HB grey lead pencils
- 6 audio tapes
- a ruler and two black pens
- a pack of playing cards
- a hair brush
- 4 highlighter pens
- a small make-up kit
- a calculator
- a pair of scissors
- liquid paper

Apart from the specification about the contents of the container, there are a couple of other constraints or limitations:

- The container should be *no larger* than 200 mm long, 150 mm wide and 80 mm high.

- The container must have at least two compartments.

The materials you use to make your mock-up will depend largely on what materials are available. Chapter 8 ('Model Making') will assist you greatly in this activity. Cardboard, thin plastic, cloth, paper, balsa wood and other craft materials may be used to produce your mock-up model.

The DESIGN BRIEF worksheet provided on page 72 can be used to develop your ideas and to record relevant information. You may also find the following worksheets useful:

- DRAWING SHEETS, pages 73-77
- PRODUCT ANALYSIS, page 78
- MATERIAL TEST, page 79
- PRODUCTION PLANNING, page 80
- PRODUCT EVALUATION, page 81
- MATERIALS COSTING, page 82
- INFORMATION RECORD, page 82 .

CHAPTER 4

COMMUNICATING YOUR IDEAS

In your technology work, it will be necessary to use drawings to help you communicate your project ideas and specifications — for your own benefit and for your teacher and anyone else involved in the production of your product. You may also need to add notes and comments to the drawing to give further details of what is required and how you are going to achieve your aims. (At the end of this book, you will find blank *drawing sheets* which may assist you in setting all this down.)

Drawings have been a part of life for thousands of years and have helped us in learning and understanding much of history. They are just as important today.

People communicate with each other in many different ways. A simple handshake is a means of communicating; so is writing letters. But without the use of *drawings* we would find it very difficult to get some kinds of ideas across to other people. Imagine trying to explain to someone that you want a seat made for your garden which has to be made from wood, steel and cement and it has to fit in between two particular trees. Without the use of a drawing it would be difficult to be precise about the dimensions or show what shape it should be.

This chapter will focus on:

- the use of drawings to assist in showing your project ideas
- different types of drawing presentations
- various drawing techniques
- the equipment needed to assist you with your drawings
- various types of drawing sheets which may be used.

Fig. 4.1 Some ideas are difficult to communicate in words alone.

Initially you may have many ideas for your project. It is advisable to put all of these down on paper — as rough sketches — as soon as you think of them. Don't wait; you may forget them. You will return to these ideas at a later date and test and evaluate each one before making a final decision on which to develop further.

In industry, the drawing and design work is generally done by a draughtsperson or designer. The drawings are then passed on to a skilled worker who uses them to manufacture the product. If the drawing is not accurate or communicates the wrong information, the final product will be made incorrectly and a lot of time and money will have been wasted. Loss of time and possible failure of your course may be the result for you if you rush your own designing or drawing, or put in insufficient research on your project.

PRESENTATION AND TECHNIQUES

The number and kinds of drawings required throughout the designing and development stages of your project will be governed by the size and purpose of your particular project and, perhaps, by the requirements of your teacher.

A simple or small product may require only one drawing to get the relevant information across, whereas a larger or more complex product may require a number of drawings. In some cases you may need to produce only *pictorial drawings* (sketches) of your ideas and then a final *production drawing*. If you are designing something for someone else to use, then you may need to produce sketches, production drawings and also *presentation drawings* to give an indication of the final appearance. (NOTE: These terms are explained in more detail below.)

Whichever presentation you use, it is important to remember to use the appropriate equipment and ensure it is in good order.

DRAWING PRESENTATION

PICTORIAL DRAWING

A pictorial drawing may be sketched, drawn with basic equipment (e.g. a ruler) or drawn with proper drawing instruments. It is generally used to give a true representation of the look and shape of the product in a 3-dimensional form. In most technology classes you will be required to use the *oblique, isometric* or *perspective* methods (see pages 22-24).

PRODUCTION DRAWING

The production drawing is the final drawing which shows all the detail for the making of the product. This drawing must show accurate and complete measurements and contain all design details. Production drawings contain various views (see page 25), planning notes, developments and layouts (see pages 25-26), as well as written information about materials (cutting lists, costings) and processes to be used on the project. The amount of detail depends on the complexity of the project. Production drawings are used by architects, builders, industrial designers and tradespeople.

PRESENTATION DRAWING

A presentation drawing is used to show the finished look of a product to any prospective client (or your teacher). It is also used to assist in advertising and marketing the product — to give the consumer an indication of the function or appearance of the product. In industry, this drawing is done by a graphic designer.

In some technology units you may be required to find a client — a friend, community group or family member — who wants you to make a product. First impressions are generally lasting, so your presentation drawing must be of a high standard. Artistic presentation requires shading, tones, colour and texture to produce an accurate overall image.

DRAWING METHODS

SKETCHING

A sketch is generally used to get down your immediate ideas and does not have to be graphically perfect. It may be done freehand or it may be improved by the use of a ruler. Your teacher may decide that a sketch is *all* that is required for a particular project. Dimensioning, materials lists, cutting lists and costing of projects can still be prepared from such drawings.

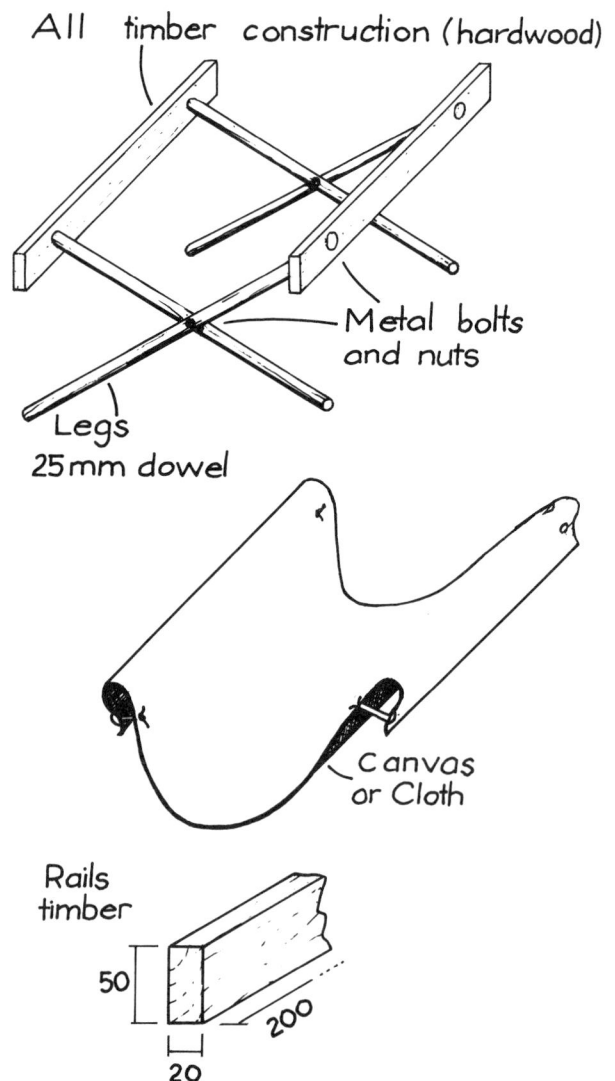

Fig. 4.2   Freehand sketches for a stool

TECHNICAL DRAWING

Technical drawing techniques are used in the production drawing — the detailed final drawing. They are usually drawn with the aid of specialised equipment such as drawing boards, technical pens and set squares (see below for more detail on equipment).

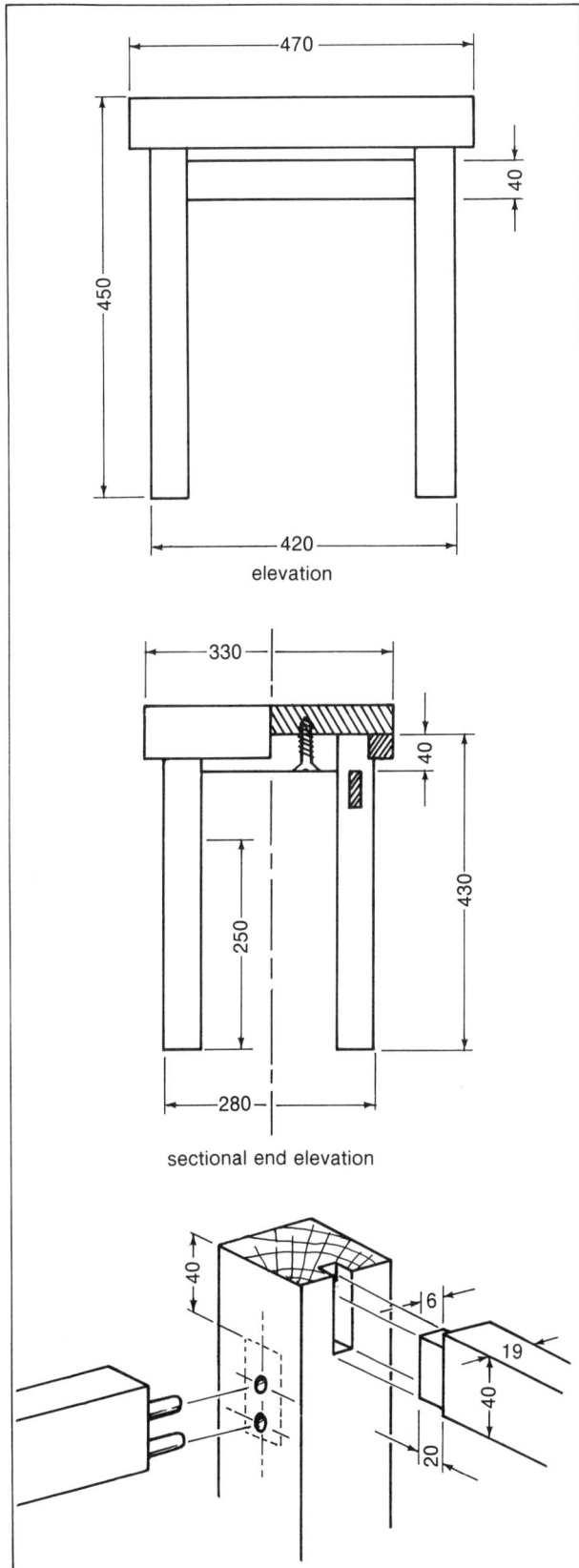

Fig. 4.3 A technical drawing

COMPUTER-AIDED DRAWING (CAD)

The most up-to-date drawing in industry today is done with the aid of computers. Designing and drawing with the assistance of the computer is certainly time-saving and easier once you have obtained the necessary skills. In CAD the designer is able to do many things such as 'turn' the object around 360 degrees to see the opposite side of a drawing. Highly complex programs enable the designer to solve problems and give answers to questions involving the actual production ('If we do X, what will happen to Y?'). The motor car industry uses CAD for this reason.

Remember, however, that the computer can only do the work correctly if it is fed correct and precise information. Programs range from the fairly basic, which may be available in your school, to the highly complicated — some can even be used to make up a miniature model of the product that has been drawn.

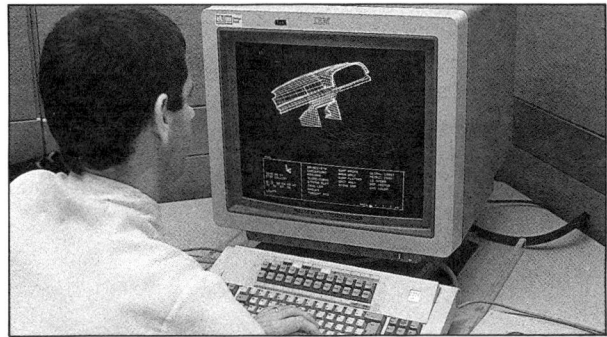

Fig. 4.4 Using computer-aided drawing (CAD) in motor vehicle design

DRAWING EQUIPMENT

The easiest and simplest method of drawing your initial design ideas is to use sketching. The equipment for this task is minimal. This is also the case when creating the drawings for your final production drawing for the project. It is only when you are required to produce a fine line drawing that you may be required to use more technical methods and equipment than those listed below.

Of course, all drawing equipment should be well maintained. Dirty and damaged equipment produces dirty and poor quality drawings.

DRAWING PENCILS

The two most used type of grey lead pencils you would use are the H and HB pencil. HB pencils, used in art classes, are useful for sketching and assisting with shading if required on technology drawings. An H pencil is a harder pencil which can be used to do fine linework and finishings on your final drawings. It is also possible to obtain these grades of pencils as clutch pencils. Always keep your pencils sharp.

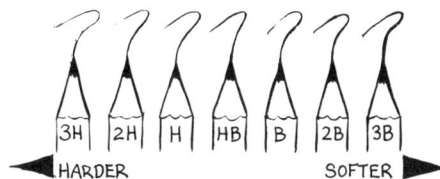

Fig. 4.5 Pencil grades

COLOURED PENCILS

Coloured pencils can make your drawings and design ideas look much better. This will assist you in the presentation of the final drawing. The writing of special notes, or the shading in of items which need highlighting can also be done with the use of coloured pencils.

ERASERS

The eraser is used to remove any mistakes you make while producing your drawings and sketches. All initial drawing should be done *lightly* in case you find it necessary to make corrections to your work. Always use good quality, soft white erasers.

Fig. 4.6   A soft white eraser

FRENCH CURVES

French curves are useful for marking curves of varying shape. They are made from plastic and are very useful when you do not have a compass handy.

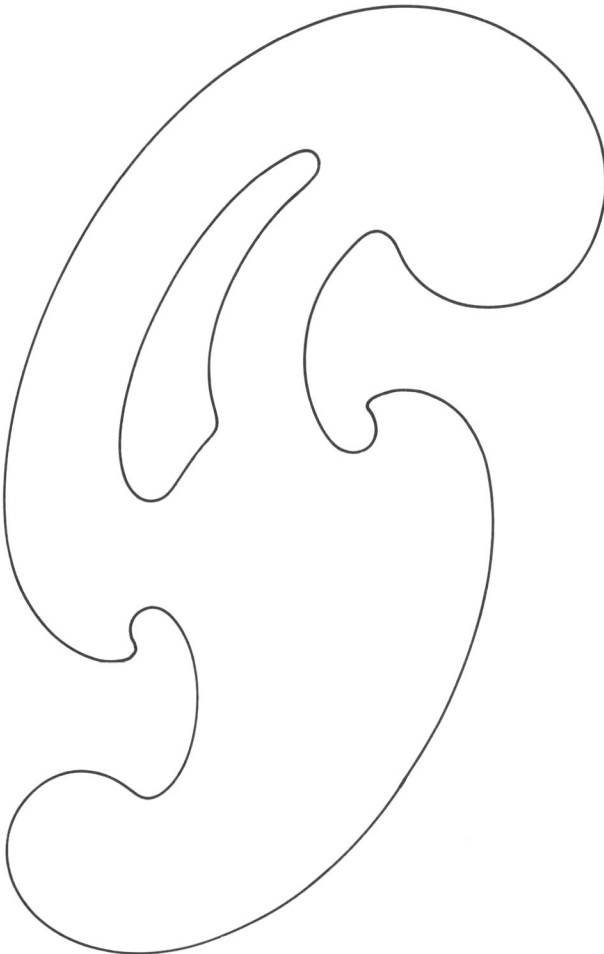

Fig. 4.7   French curves

COMPASS

The compass is used to mark out circles, arcs, bisect lines and so on, and can also be used to mark out measurements. They are usually made from metal.

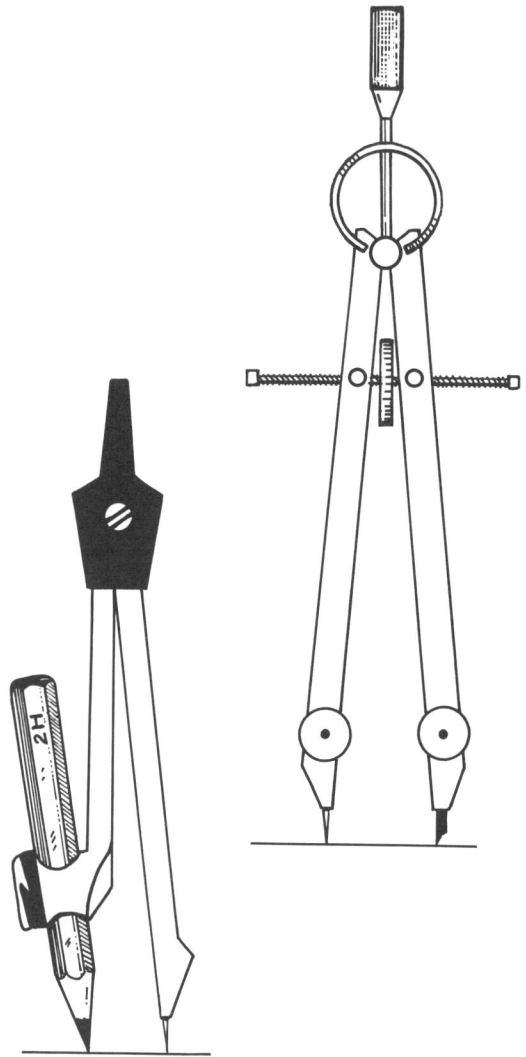

Fig. 4.8   Compass and spring bow

SET SQUARES

Set squares are used to draw in lines at specific angles. They can be obtained with various angles. The two you will use for your oblique and isometric drawings are the 45° and 60°–30° set squares.

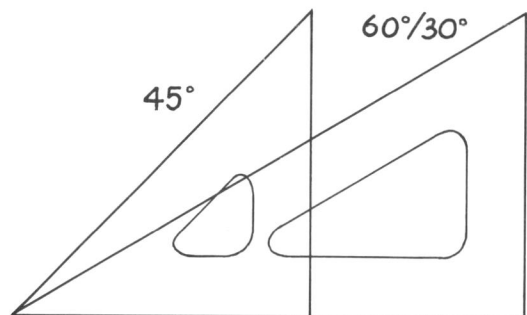

Fig. 4.9   Set-squares

DRAWING BOARDS

Drawing boards can be bought or made. They should have a slight incline. Drawing boards are used essentially for the final drawing, especially if you are required to produce a presentation drawing for assessment purposes. (Your initial drawings may be sketched or drawn by just sticking your drawing sheet on a flat board.) Spring clips or masking tape can be used to hold your sheets on the board.

T-SQUARE

The T-square is used in conjunction with the drawing board. It can be used to draw horizontal lines and to help guide the set square when that is used. Always look after the T-square. It must be at 90° at all times, otherwise your line work will not be at the required angle and this will cause problems when constructing your drawings.

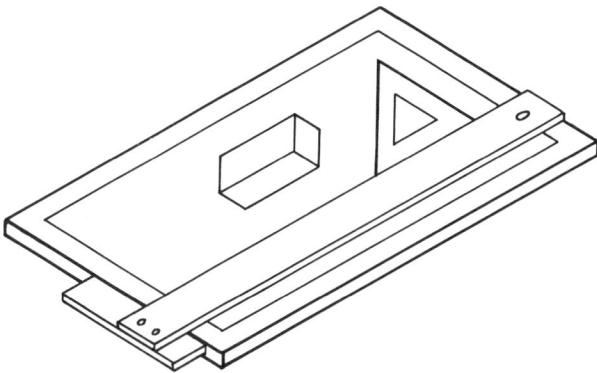

Fig. 4.10   A 45° set-square, T-square and drawing board

PAPER

If your teacher has not already made up a special drawing sheet for you to use, you will probably be able to select from normal white paper, graph paper or isometric grid paper. In some cases you may need to use all of these. It will depend on the type of drawings you require. Paper comes in various sizes but you would normally use either A4 (297 mm X 210 mm) or A3 (420 mm X 297 mm).

MARKING PENS

If you are fairly proficient in your drawing skills you may wish to use marking pens to finish off your line work. The fine types — 0.1 or 0.2 — are the most commonly used. The use of marking pens is not easy and requires care. It is advisable to ask your teacher about the use of them first.

OTHER EQUIPMENT

There are a number of other drawing aids available from graphics and art equipment suppliers. A few are listed below to give you an idea of the variety.

ERASER GUIDE – helps erase lines without touching others.

TECHNICAL PENS – ink pens used by designers and experienced drawing people to do fine quality linework.

FLEXICURVE – a piece of flexible rubber used for drawing unusual and oddly shaped curves.

PROTRACTOR – used for marking angles from 0° to 180°

TEMPLATE – a piece of flat plastic with various shapes cut out of it. Templates vary in both shape and size.

DRAWING TECHNIQUES

LINES, SHAPES AND FORMS

At first, your drawing skills may be limited but once you learn about the basic construction of lines you will be able to produce a drawing of reasonable standard. As you continue to practise, you will become more proficient and find this type of drawing enjoyable.

LINES – All drawings are made up of a series of lines which are used to build up a required shape or form. Lines can be added to show information or to improve appearance. The first thing you need to know is how to use lines to achieve this.

The simplest line is the straight line. When drawing straight lines move the pencil away from the centre of your body (see Fig. 4.11a).

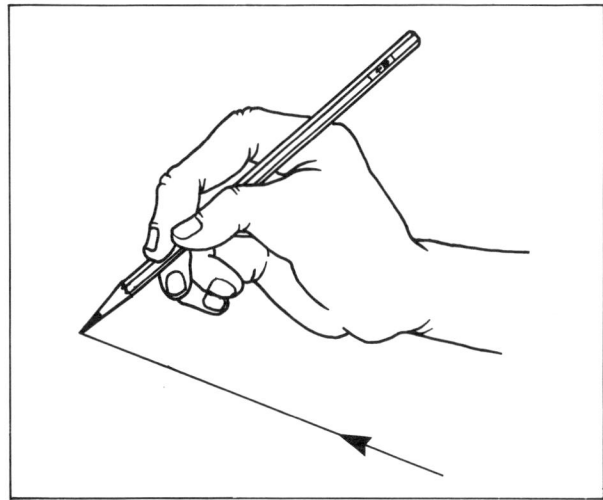

Fig. 4.11 a

SHAPES – Drawings that have only *two dimensions* are called shapes. The two dimensions are length and width.

SHAPES

square

triangle

circle

octagon

Fig. 4.11 b

FORMS – Forms have depth as well as length and width — that is, they are *three dimensional*. The depth line creates the third dimension (see Fig. 4.11c).

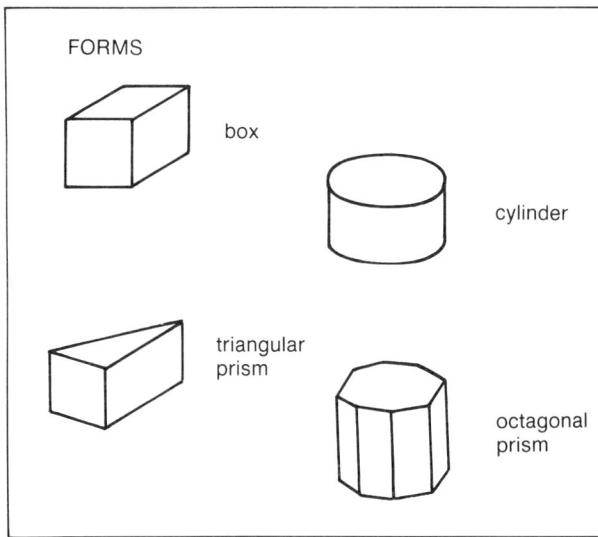

Fig. 4.11 c

OBLIQUE, ISOMETRIC AND PERSPECTIVE DRAWING

The three techniques for pictorial drawing you need to be aware of are the oblique, isometric and perspective. Each of these have their advantages and disadvantages. Your teacher will assist you to decide which method is most suitable for your particular task. Pictorial drawing will assist in giving you and your teacher or client an indication of a product's eventual appearance. A more detailed drawing may be required if your project is fairly complicated.

OBLIQUE DRAWING

This is possibly the easiest method for you to use when sketching your products. Firstly, you must draw the shape of one side of the object — the 'true' shape. This is viewed from straight on, as in Fig. 4.12b, and uses only horizontal and vertical lines (two dimensions). The third dimension is created by drawing the depth line at 45° to the horizontal line. (NOTE: If you are drawing odd or cylindrical shapes it is often easier to draw a cube first and sketch the shape from there.)

The process of oblique drawing is shown in Figs 4.12b to 4.12e.

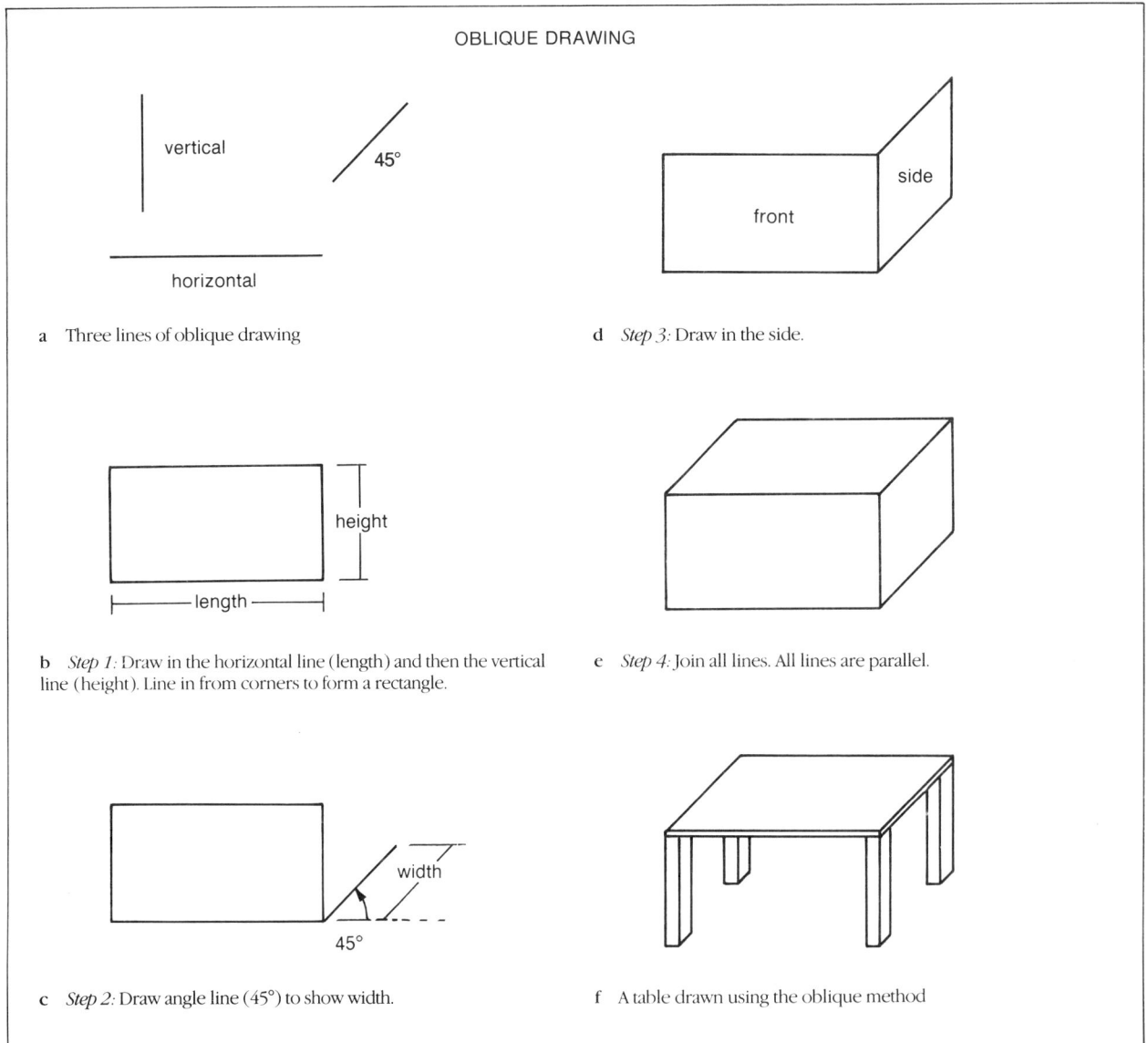

OBLIQUE DRAWING

a   Three lines of oblique drawing

b   *Step 1:* Draw in the horizontal line (length) and then the vertical line (height). Line in from corners to form a rectangle.

c   *Step 2:* Draw angle line (45°) to show width.

d   *Step 3:* Draw in the side.

e   *Step 4:* Join all lines. All lines are parallel.

f   A table drawn using the oblique method

Fig. 4.12

ISOMETRIC DRAWING

Isometric drawing is similar to oblique drawing. Although it is not as easy to produce, it does look more realistic than oblique drawing. Isometric drawings have a vertical height line (the 'corner') which is 90° to the horizontal line, and length and width lines which are drawn at a 30° angle from the horizontal line (see Fig. 4.13a).

The sequence of construction is shown in Figs 4.13b to 4.13e.

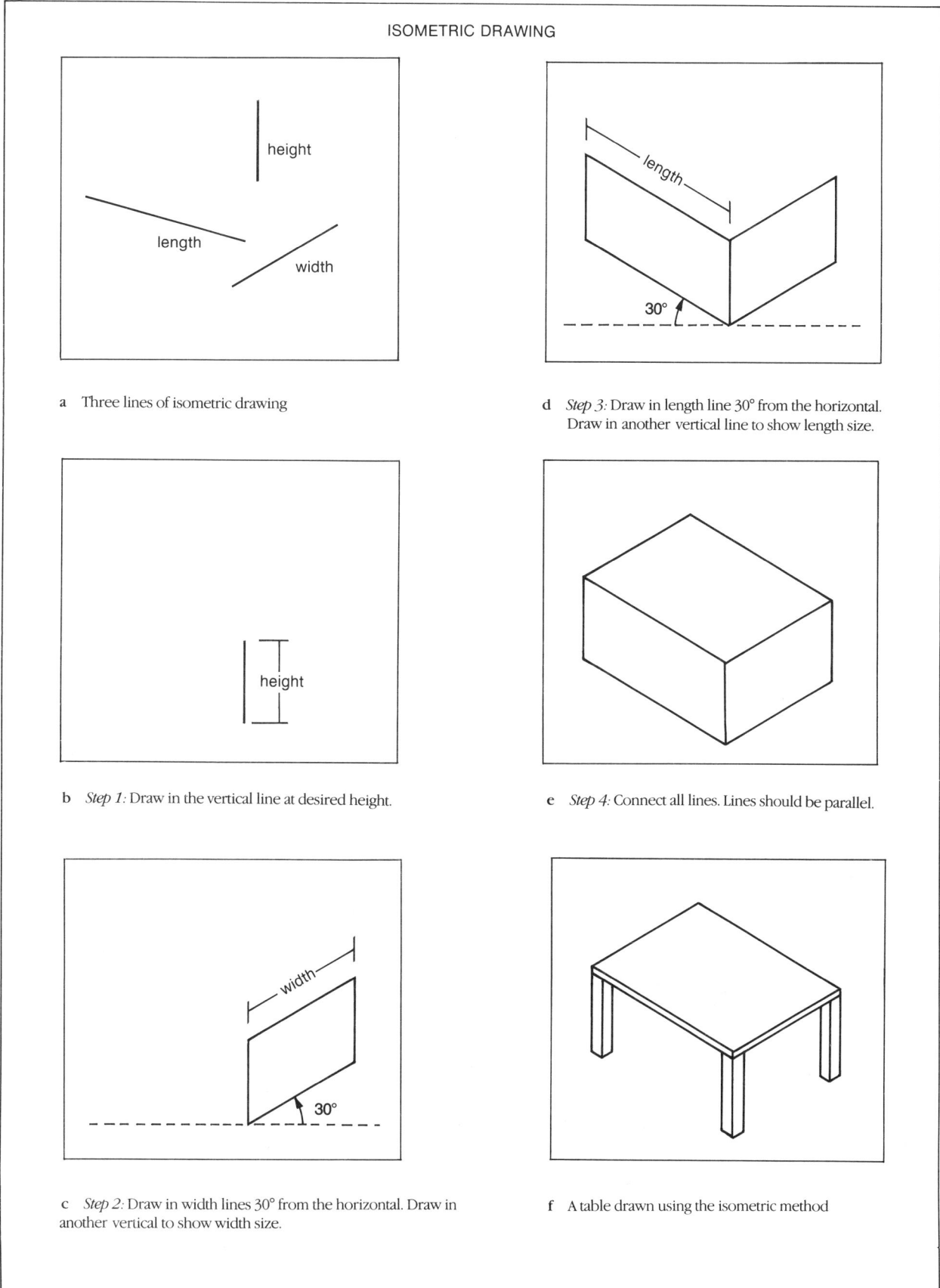

ISOMETRIC DRAWING

a   Three lines of isometric drawing

d   *Step 3:* Draw in length line 30° from the horizontal. Draw in another vertical line to show length size.

b   *Step 1:* Draw in the vertical line at desired height.

e   *Step 4:* Connect all lines. Lines should be parallel.

c   *Step 2:* Draw in width lines 30° from the horizontal. Draw in another vertical to show width size.

f   A table drawn using the isometric method

Fig. 4.13

PERSPECTIVE DRAWING

Perspective drawing is an effective method of showing 3-dimensional objects. In *one-point perspective*, the object is viewed from the front (true shape) and all lines from the corner are then drawn back to a selected point on the horizon. This is called the vanishing point.

The sequence for the construction of a rectangle in one-point perspective is shown in Figs 4.14a to 4.14c.

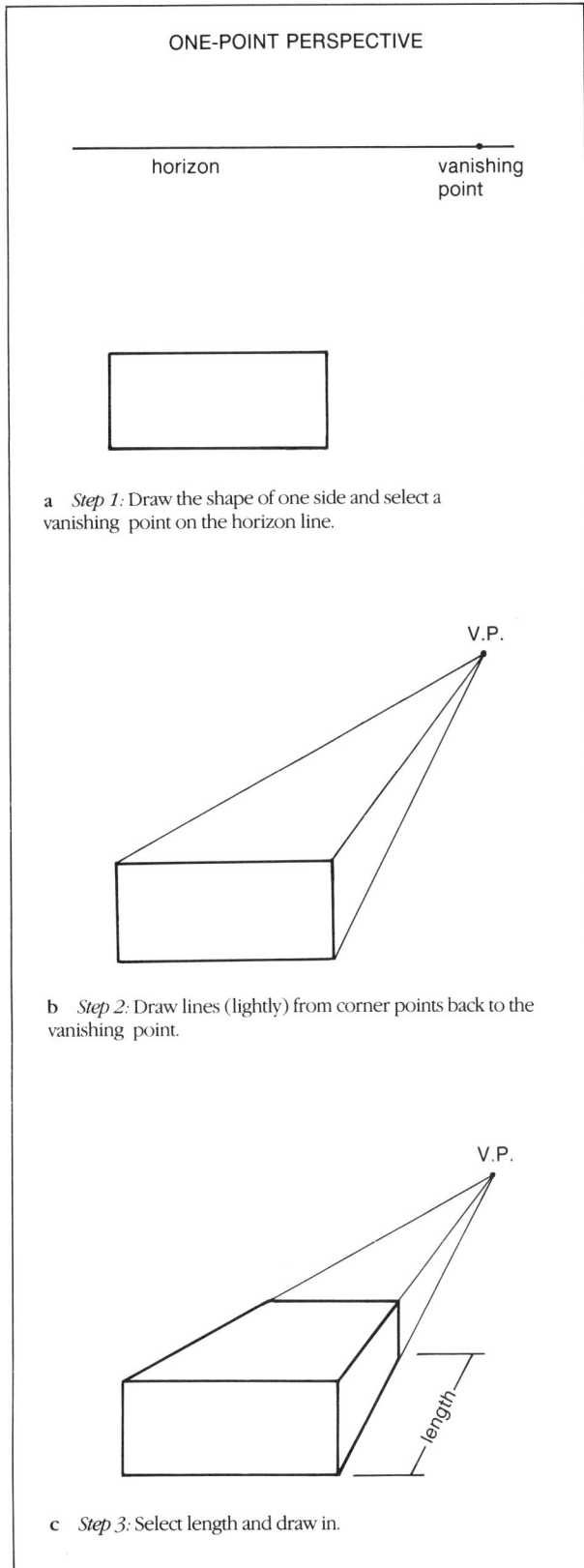

In *two-point perspective* drawing (see Figs 4.15a-d), two vanishing points are used and the image viewed from the nearest edge. The vanishing points are placed apart along the horizon line. An estimation of the height is made and the lines are then drawn back to the two vanishing points. The back vertical lines are drawn in and then the top lines are drawn. All lines should be drawn *lightly* at first until you are sure of the drawing being correct.

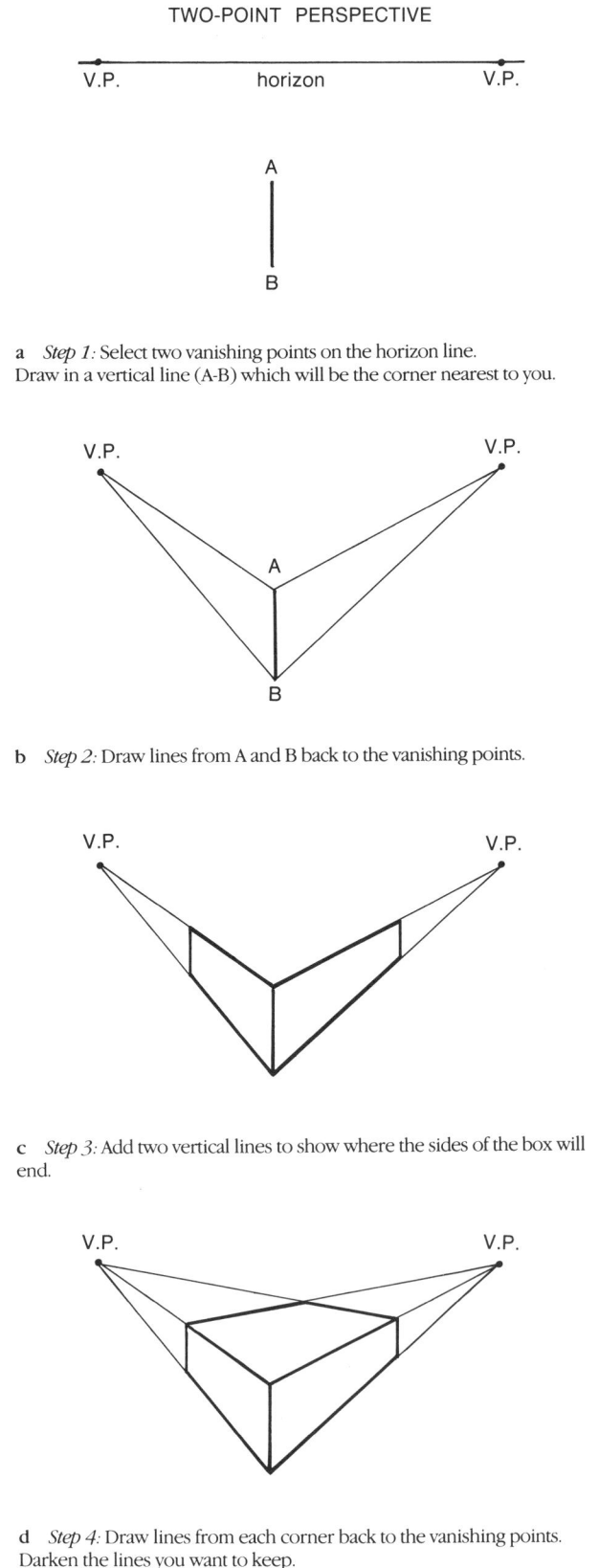

ONE-POINT PERSPECTIVE

a   *Step 1:* Draw the shape of one side and select a vanishing point on the horizon line.

b   *Step 2:* Draw lines (lightly) from corner points back to the vanishing point.

c   *Step 3:* Select length and draw in.

Fig. 4.14

TWO-POINT PERSPECTIVE

a   *Step 1:* Select two vanishing points on the horizon line. Draw in a vertical line (A-B) which will be the corner nearest to you.

b   *Step 2:* Draw lines from A and B back to the vanishing points.

c   *Step 3:* Add two vertical lines to show where the sides of the box will end.

d   *Step 4:* Draw lines from each corner back to the vanishing points. Darken the lines you want to keep.

Fig. 4.15

OTHER DRAWING TECHNIQUES

There are a number of other drawing techniques that you could use to draw your projects. The most suitable for your needs are those already described. A few are listed below:

● axonametric projection
● orthographic projection
● centre line method
● grid or box method.

VIEWS

Views are slightly different from other drawings in that they tend to show much more detail of the inner parts of an object. The two views that are most common are the sectional view and the exploded view (see Figs 4.16a and b).

SECTIONAL VIEW

This is a view which enables you to see what the internal parts of an object (or its components) look like when 'cut' through a section. Sectional views are good for showing hidden or unseen areas.

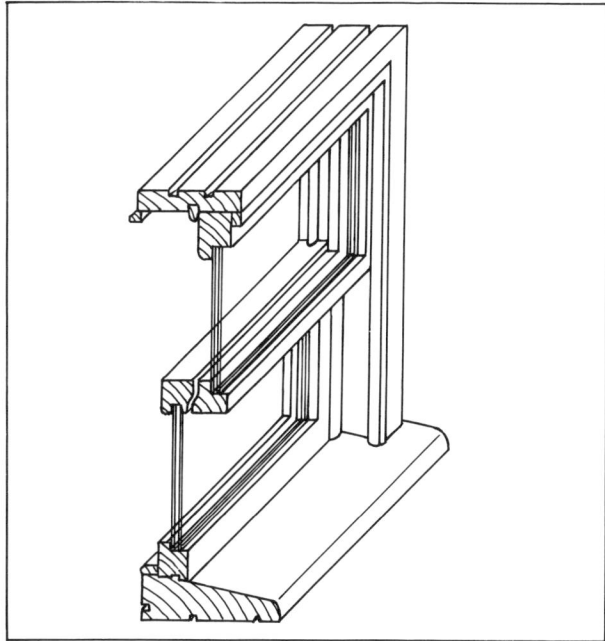

Fig. 4.16 a A sectional view of a window frame

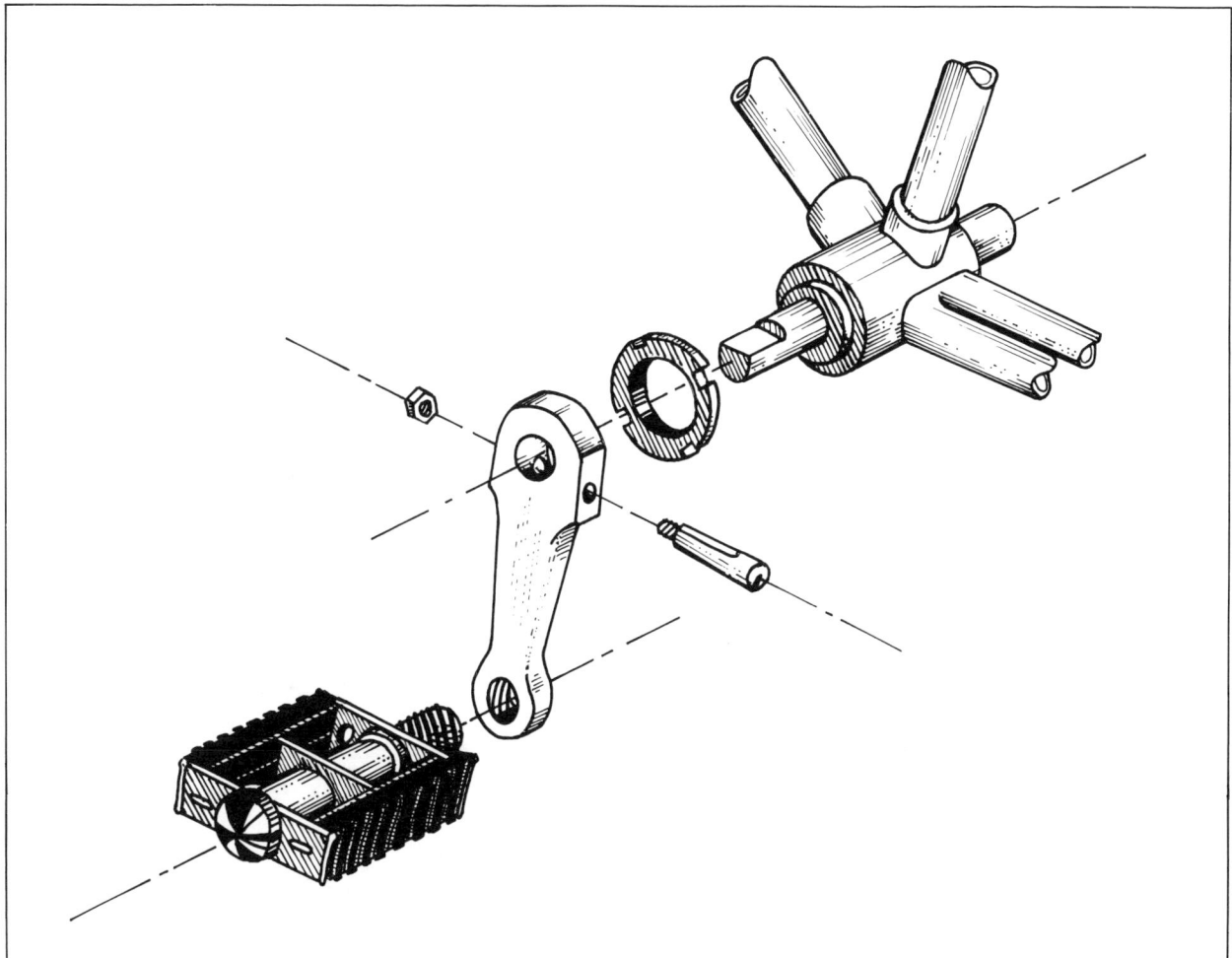

b An exploded view of a bicycle pedal

EXPLODED VIEW

This view shows each part individually. The advantage of this type of view is that you can visualise each part which makes up the object, and where and how the parts all fit together. Exploded views also help when you need to represent notes and detailed information about the object. This is very useful when making a product (such as a nest of drawers or a tool cabinet) which has many parts, or complicated structures or joining methods.

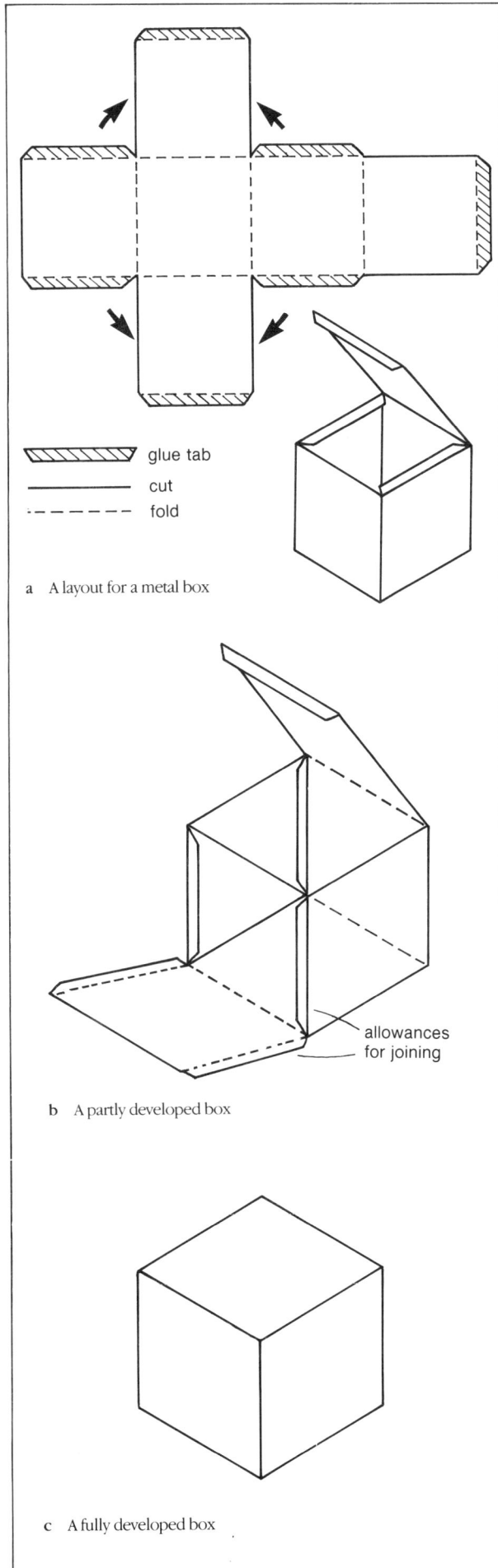

glue tab
cut
fold

a   A layout for a metal box

allowances for joining

b   A partly developed box

c   A fully developed box

Fig. 4.17

DEVELOPMENTS AND LAYOUTS

When designing and drawing products which are being made from a pattern or flat material such as metal plastic, cardboard or fabric, the item must be drawn in its *flat* state and all fold lines and allowances for joining must be indicated accurately. A layout shows the item in its flat state (Fig. 4.17a) and a development shows how it will look during construction (Figs 4.17 b and c).

DRAWING SHEETS

The style of drawing sheet you use for designing your products will largely depend on what is required in your particular course. Some courses require you to write down the materials to be used and to cost the total project; others may want cutting lists drawn up, as well as indications of equipment and tools to be used.

It doesn't matter what type of drawing sheet you use as long as it addresses the needs of the course work you are doing and is presented in a neat and orderly manner. It is good practice to put borders around your drawings and include your name and the project name. Some DRAWING SHEETS are provided on pages 73-77.

REVISION QUESTIONS

1   What is the main reason for using drawings?

2   Name *three* types of pictorial drawings you may use in designing and drawing your product.

3   Circle TRUE or FALSE:

   a   An oblique drawing has a vertical, horizontal and depth line.   T   F

   b   One-point perspective has two vanishing points.   T   F

   c   A 2B pencil is harder that a 2H pencil.   T   F

   d   Sectional views show each part of an object separately.   T   F

4   What is meant by '3-dimensional drawing'?

5   What is the purpose of a presentation drawing?

6   Explain the difference between a sectional view and an exploded view.

STUDENT ACTIVITIES

1   Select *one* of the items below and complete a sketch of it in both oblique and isometric form.

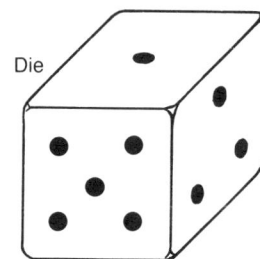

Die

Matchbox

FLASH

2   Draw a layout for the construction of the carton shown below. Note that all corners have a 10 mm tab.

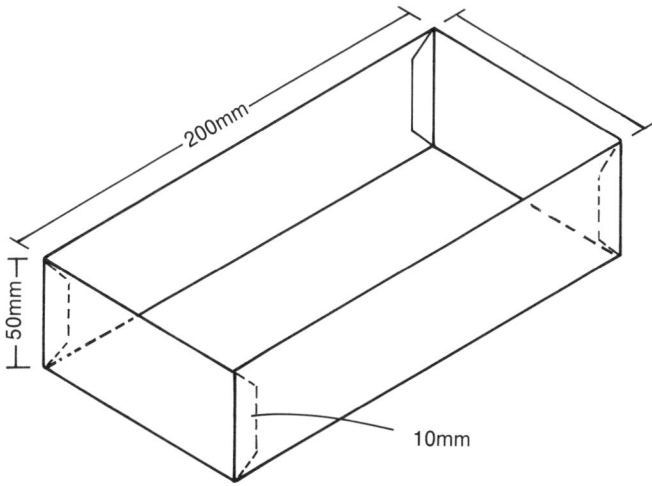

3   Draw a vertical sectional view of the house brick along the 'cut' indicated by the dotted line.

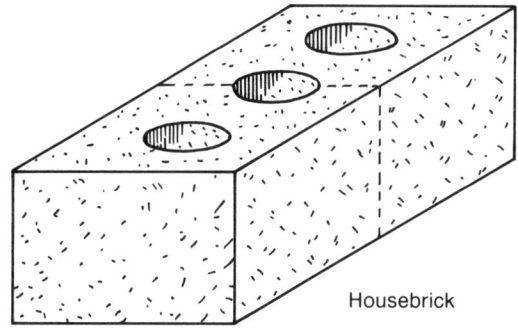

200mm

50mm

10mm

Housebrick

CHAPTER 5

APPROACHING DESIGN

The way things look and function, and how much they cost are important factors that consumers take into consideration when buying products. These and many other aspects are always in the mind of the designers who develop items for the consumer market. In this chapter you will learn more about these and other aspects of design.

At the end of the chapter are some design exercises which will give you further practice with design briefs and the development, analysis and evaluation of products.

This chapter will focus on:

● what makes up design

● how design affects our selection of products

● important aspects to be considered when designing

● product analysis.

We are all designers in some way, whether it be in laying out our own bedrooms, making a Billy cart or Cubby House, choosing a hairstyle, renovating the house or just in altering the colour and accessories on our bikes. Whenever we make new things or change the old we are required to start thinking about all of the important aspects of good design.

Most of your design efforts would have been made by trial and error. However, as you became more experienced and knowledgeable through learning from your mistakes, making and designing would have become easier. You may also have found that by working through your needs *systematically*, things were easier to do. It is no different in the manufacturing industry. It is essential for manufacturers and industrial designers to work systematically and make sure they have thoroughly investigated all aspects when designing and making products for consumer use.

As you have learnt from earlier chapters, drawings and sketches are a large part of designing. They are used as the communication tool between the designer and the client — the manufacturer or the consumer. The design has to take into account a number of factors. These elements need considerable thought before decisions can be made about the viability of the final product.

The designer needs to eliminate all problems before the product is put on the market or it may fail to meet

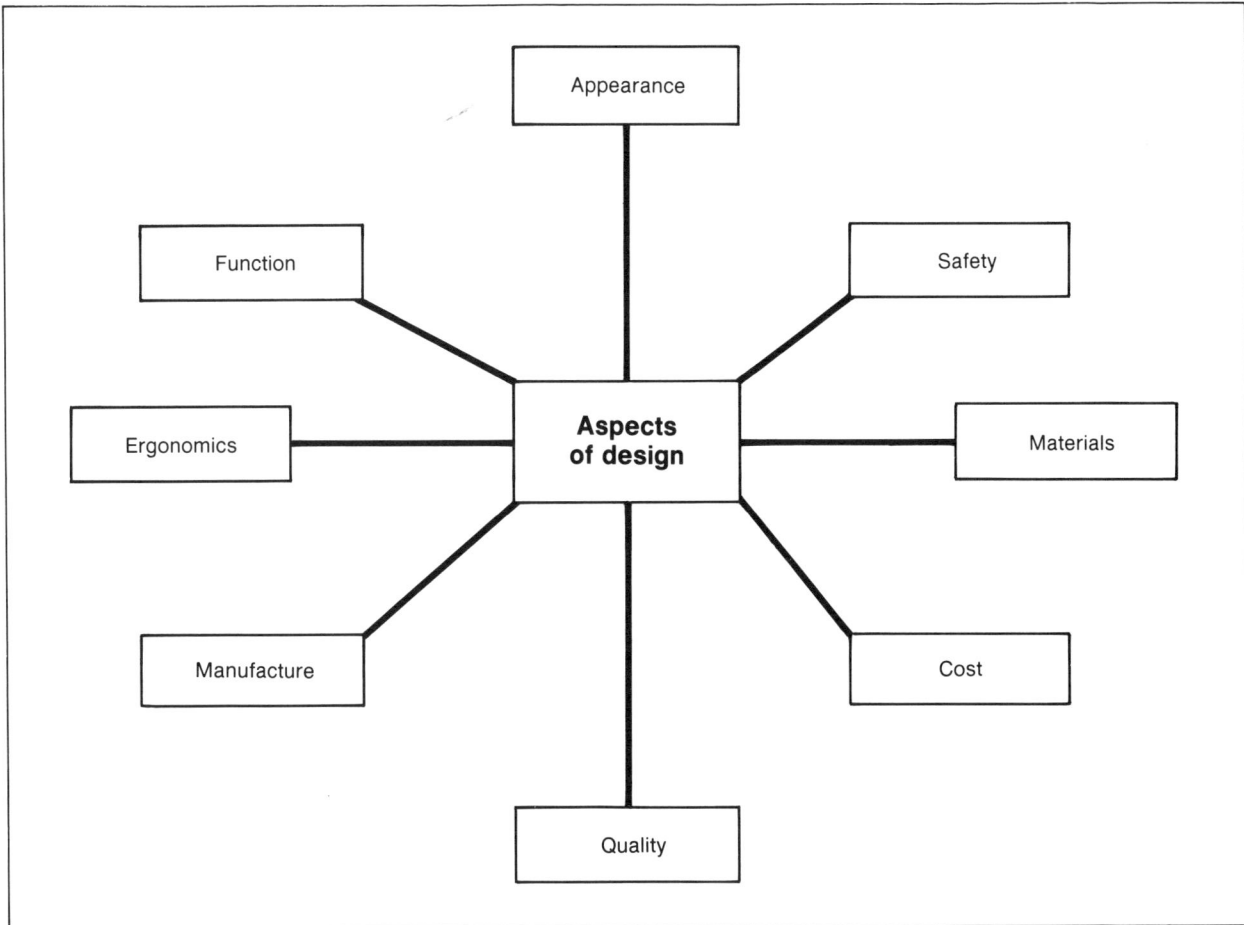

Fig. 5.1

required standards and may be recalled. For certain products, such as toys, automobiles and electrical appliances, there are regulations which govern the standards of products. If the standard isn't met a product will not be released on to the market until it is fixed. This is an expensive process and could easily ruin the chances of the product being successful in the market place.

The same need for care and attention to detail applies in respect of your own projects. You need to be sure your product carries out satisfactorily the function for which it was designed.

ASPECTS OF DESIGN

When designing, you (the designer) need to take into consideration many of the aspects listed in Fig. 5.1. These are explained in detail below.

FUNCTION

The original intention of the project must be kept in mind at all times. If the product doesn't meet the initial criteria set down in the design brief (see Chapter 3), then it could be said to be an unsuccessful design. It is not much good designing a cassette rack that unbalances and tips the cassettes out, or a knapsack that has pockets which are too small to hold the books you intend it to carry. Problems such as these can occur through lack of research or relevant information in the design brief.

The manner in which the product is to be used also determines many of the decisions made in the design

stage of your project. Many questions need to be answered. Here are a few for starters:

- What are the essential features of this product?
- Where is the product to be used?
- Is it for outdoor or interior use?
- Does it need to hold weight?
- Is it environmentally safe?
- Will objects (e.g. hot drinks) be placed on it?
- Do special safety standards need to be met?
- Is the product's size important?
- Can it be made with the available tools and equipment?
- How does it operate?

Fig. 5.2   What is wrong with the design of this chair?

COST

The cost of a product is an important concern for all involved. If the product is too expensive consumers will generally not buy it. On the other hand, to keep costs down, manufacturers may be tempted to cut corners and produce an inferior product. But cost should always be weighed against the quality of the product. Unless this is done you cannot gain a fair indication of a product's real value and usefulness. Many companies stake their reputation on their products and would never consider trading cost against quality. These companies are often the very successful ones.

Cost factors may also affect your own class productions. The school department will have a budget and this usually determines the amount of money each student is allowed to spend on materials and accessories for their projects.

Two other factors need to be considered when assessing cost:

TIME – In industry, 'time is money'; that is, the cost of labour (wages and salaries) has to be taken into account in the design *and* manufacturing stages. In your classwork, this will not be a real cost because your labour will not be measured in dollars. It may, however, be measured against time spent on other work — so don't waste it.

REPAIR AND MAINTENANCE – The likely costs of repairs and maintenance of a product (parts and time) will be important to the buyer and need to be considered in the design and manufacturing stages.

ERGONOMICS

Designing products for people to use is very tricky and takes considerable research and thought. All human beings have the same basic form. However, we all have our own peculiarities, such as height. Ergonomics, a relatively new science, is the study of the relationship between people and their physical working environment. Its findings help designers to design products (e.g. adjustable furniture) to take into account the *differences*

in people. (Surveys are a good source of information to be used for this aspect. There is more about surveys on page 55.)

Of course some people have very special needs and require custom-made products. This could be an artificial limb for a person who has lost an arm or leg in an accident, or a precision-sized bike for an Olympic cyclist.

Other factors that need considering when designing for the human form are as follows:

- Comfort
- Safety
- Strength (of the user)
- Grip (of the product)
- Stability (of the product)
- Mobility (of the user)
- Weight and height (of the user)

Here are some questions that could be asked:

- Who will be using the product?
  – a tall or short adult?
  – a big or small adult?
  – a teenager or small child?
  – a disabled person?
  – an elderly citizen or baby?
- What is it used for?
  – to store things?
  – to carry things
  – for sitting in?
  – to assist my exercise program?
- Under what conditions will it be used ?
- Is it easy to use?
- Are important switches and knobs clearly marked?
- Can it be opened easily?
- Do you want young children to be able to open it?
- Are there safety problems to consider?

Fig. 5.3   People differ in size and shape. Differences need to be taken into consideration.

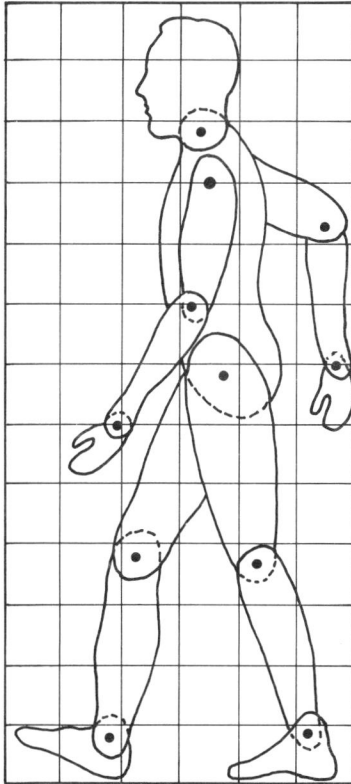

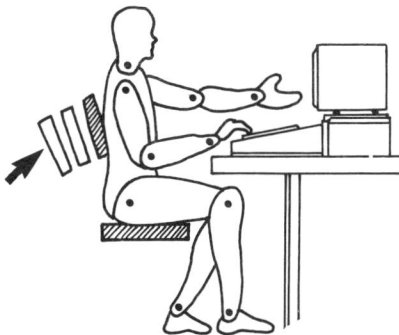

Fig. 5.4 The ergonome model is very useful for showing relative limb movement.

MATERIALS

The choice of materials used in manufacturing products depends largely on what the product is to be used for. If the product is for outdoor use, the material will need to have different properties (specific qualities or abilities) from a material which is to be used indoors. If it needs to be light in weight, it might be made from plastic or aluminium.

Today there are 'super' materials such as carbon fibres which are light but strong and thin like hair fibres. These are used for strengthening other materials. Materials engineers are often asked to invent *new* materials for specific jobs. The world of materials is vast and complex and it takes much experience and knowledge to be able to produce and decide on the best type of material to use in making products for consumer use.

Fortunately for you, the range of materials you may be able to choose from is fairly limited. This does not mean, however, that you need not be careful in your selection of materials. (For further information on this refer to Chapter 6.)

PROPERTIES OF MATERIALS

Each material is said to have 'properties'. These are its special characteristics, qualities or abilities. For example: aluminium is *light*; denim is *tough*; copper is *malleable* (easy to work); pine is *easy to cut*. The properties of the materials need be considered at all times. Here are some questions which may help with this task:

- What is the function of the project?
- What tools and equipment are needed to form the selected material?
- How can I join the material?
- What are its advantages?
- Are there any disadvantages?
- What must I be careful of when using this material?
- What will I need to cut the material?
- Is the material harmful to health or the environment?

Some of the possible properties to look for are:

- softness/hardness
- aesthetics (beauty, e.g. natural finish)
- resistance/conductivity
- stiffness
- toughness
- strength
- weight
- feel
- colour

HEALTH AND SAFETY

The product must always be safe for use, otherwise the product could be forced off the market. All States have health and safety regulations, and consumer protection organisations keep a check on health and safety factors as well as other consumer complaints. In your class, the teacher is responsible for this so make sure you take notice of and follow the safety guidelines provided by your teacher.

Fig. 5.5 Safety is a special consideration in some products.

APPEARANCE

Appearance depends on such elements as the colour and, texture of the material and the shape of the product. It is important not to judge a product on appearance alone without considering how well it functions. The best product is one that combines attractive appearance with functional quality. You will be required to design and finish off your products to a high standard because of assessment but, hopefully, you will also do so for reasons of personal pride.

In some technology courses you may be required to work for a client and the consideration of appearance will be most important to remember.

MANUFACTURE

The methods used to manufacture your product and the quality of the finished item need to be taken into consideration at the designing stage. It is essential that you make sure that all the tools and equipment you require are operational and available, otherwise you will spend many hours redesigning at a later date. It is much easier to come up with alternative methods at the initial stage.

The quality of the end product often depends on the way it is manufactured. If you rush your work or use incorrect methods, the quality will suffer. Consideration also needs to be given to whether any of the manufacturing techniques and processes you use are harmful to the environment or the health of the person using them (in this case, you) — and the value of the product when weighed up against the use of natural resources (gas, coal/electricity) to operate the equipment to produce the articles.

Fig. 5.6   Use of poor quality materials may result in unsafe products.

Here are some questions to be asked about manufacturing:

- Are there better and less dangerous methods of carrying out this work?
- Is the method used the most environmentally acceptable?
- Is the method used the most economical?
- What skills and knowledge do I need to operate the equipment?
- Do I require special guards or screens around some types of machinery and equipment? If so which ones?
- Are there any special regulations and codes of practice I should follow?

QUALITY

The real test of a product comes when the product is completed and being used. If the quality is poor, the product will function badly. Poor joining processes may cause chairs, tables and ladders to collapse; plastic, metal or fabric products may come apart at the seams.

Safety is closely associated with quality of work, and products need to be checked thoroughly before being used — especially those that will be placed under forms of stress or will be used by children. Quality control is the responsibility of the maker.

DESIGN EXERCISES

The following design exercises will assist you to gain:

- further experience in the process which is used by designers to solve problems when designing products for consumers;
- experience in product analysis and evaluation.

The worksheets listed on page 16 will be useful.

PROBLEM SOLVING

Select *one or more* of the design exercises from the nine listed below and complete the task to the requirements outlined. Follow the problem-solving process outlined in Chapter 3. All exercises should contain:

- drawings
- a materials list
- a costing list
- a production sequence detailing, step by step, the procedures used.

1   Design a piece of outdoor furniture for a garden. Use a picture of a real garden (photograph, magazine cut-out) and make sure the furniture will blend with the setting.

2   Design a front fence to suit a house you have selected from the real estate pages of the local paper. The fence is to be no higher than 1400 mm. Material selection is your choice. Don't forget the foundation. Happy fencing!

3   The local council is running a competition which involves school students. The competition is to see who can design the best-looking, most functional and cost-efficient piece of playground equipment possible. The piece of equipment must be suitable for

use by a wide age and ability level of children. This is a real test of your imagination. Have a go and see how creative you can be.

4 Select *one* of the groups below, then design and produce a packaging system which will pack the items safely, be functional and assist in the marketing of the product — that is, be inviting to the consumer.

GROUP 1	GROUP 2
1 claw hammer	2 lipsticks
1 4-piece screwdriver set	1 blusher set
1 300 mm ruler	1 hair brush
1 tape measure	1 comb
3 chisels	1 mascara brush

You can decide the type and size of items to be stored.

5 A number of special occasions and situations have been listed below. Select *one* and design a suitable outfit for the occasion. Make up paper patterns to show your ideas and designs.

● your sister's wedding

● your 21st birthday

● Melbourne Cup

● a fancy dress party (1920s theme)

● Logies Award night

● exercise classes.

6 A new football team from Tasmania has been admitted into the Australian Football League. Design a suitable logo (emblem) and jumper to be used by the new team.

7 The school is organising a Billy Cart design competition and race for the end of term. The competition will be judged on the three criteria listed below. Design and make a prototype of the winning Billy Cart.

CRITERIA	CONSIDERATIONS
Aesthetics	appearance
	materials
Design	aerodynamics
	materials
	ergonomics
Function	speed
	handling
	steering mechanism

8 Design a wooden jigsaw puzzle suitable for a child aged six to ten years. The puzzle should fit inside a cardboard storage box measuring 250 mm x 300 mm x 20 mm.

9 Design and make a prototype of a mobile storage unit to be used to house items from *one* of the following groups:

GROUP 1	GROUP 2
a television set	hand power drill
a video recorder/player	power jigsaw
10 video cassettes	claw hammer
slide projector	tape measure
	small hand plane
	6-piece screwdriver set
	tenon saw
	2 G-clamps
	attached vice

PRODUCT ANALYSIS

From the group of products listed below, select one that you would like to analyse and evaluate. Assess it in terms of its function, appeal, quality and so on. Refer to Fig. 5.1 for the full list of aspects to consider. Use the PRODUCT ANALYSIS sheet on page 78 to record your findings. This will make your task much easier.

● Sanyo 43 cm television set
● Lee jeans
● Remington shaver
● Holden family car
● a Sidchrome spanner
● Victa mower
● a Gunn and Moore cricket bat
● Wella Balsam shampoo
● a leather handbag
● a piece of metal furniture
● Hills clothes hoist
● electric knife

You may find it interesting to select two similar products and do a *comparative* analysis — for example, compare a leather handbag with a vinyl handbag, or a 1990s telephone with an early 1970s model.

REVISION QUESTIONS

1 Name the eight important aspects of design mentioned in this chapter.

2 What is meant by the description 'ergonomically designed'?

3 Circle TRUE or FALSE:

a Design is mainly concerned with the appearance of products. T F

b Design aspects are often used to assist in the sale of products. T F

c Ergonomics is the study of the economy. T F

d A graphic designer designs products such as household goods and cars. T F

4 Why is it important to understand about the properties of materials being used in a product?

5 Look at the products listed in Design Exercise 9. Which of these were designed especially with safety in mind?

6 Explain why there is a need for the testing and evaluating of products once they have been designed?

STUDENT ACTIVITY

Research *two* of the items listed below and give (for each) an example of design changes in the last ten years or so.

● tennis racket

● surf board

● motor car body

● playground slide

● bicycle frame

CHAPTER 6

MATERIALS

Before products are made, there are a number of steps which need to be taken to make sure that the materials used are appropriate for the purpose. This can be achieved by testing and evaluating the materials against the product requirements. (Clay would certainly *not* be suitable for making a basketball, but it isn't always so obvious.) Testing can also be useful to assess various materials by comparing their capabilities with other, similar, materials. In most cases, the *best* material would be used. However, its availability and, in some circumstances, its cost may have to be taken into account and an inferior material may be used in preference.

This chapter will focus on:

● the historical use and development of materials

● how knowledge of the properties of materials assists in the selection of materials and processes

● the process used in selecting materials for specific purposes.

The ability to develop and use the materials around us has played a big part in improving the quality and style of living of human beings.

The first materials the human race came in contact with were those in the natural environment. Rocks were used as weapons to kill prey or to fight off raging animals and invading tribes. Wood was used for fires to cook food, provide warmth and security and, later, to smelt minerals to form new metals. Stone and wood were used to make spears to kill animals for food and their skins then used to make clothes. When not able to live in caves, people had to make shelters from the branches and bark of trees. Trees were used in making canoes for fishing and transport. These small but significant steps in using available materials helped people to improve their existence and were early forms of what we now know as technology.

When we take a look back through the ages, we can see how the number of available materials increased as the years passed by. Many early cultures are identified by the dominance of a particular material at various stages of their development, for example, the Stone Age of Europe and the Bronze Age of Greece. As people settled down in larger, more organised groups, their needs became more complex. Through experimentation, a wider range of materials — and variations on these — became available.

Today, we have thousands of material variations to choose from: 'super' metals, alloys, composite materials, treated woods, numerous plastics, ceramics and super fibres.

As long as the human race has a need or desire to improve its quality of life, the development of materials will continue. The knowledge we gain when working on this development will be important and we need to weigh up the overall benefit against any problems the development may cause. (The dilemma of technology has always been with us: a tree branch can be used to make a dwelling or it can be used as a weapon.) It is essential to take into consideration the effect of any development on society, the economy and the environment.

SELECTION OF MATERIALS

The selection of the correct material for your own projects is of utmost importance. When making decisions about the suitability of materials, you need to keep in mind not only the main material being used but also the materials used for any components (parts) or accessories.

Engineers, industrial designers and tradespeople are always having to make decisions concerning the selection of the best material for a particular need. If they select the wrong material, the product may not carry out the function for which it was produced.

Fig. 6.1 A fishing line and rod must have high tensile strength.

Remember that wood, metal, plastic and fabrics are just broad names for large groups of materials. There are many hundreds of varieties in each group and all have different properties. These properties should be kept in mind when making decisions about the suitability of the material for your product or component. Selection of an incorrect material could very well affect the quality and function of the product you are making.

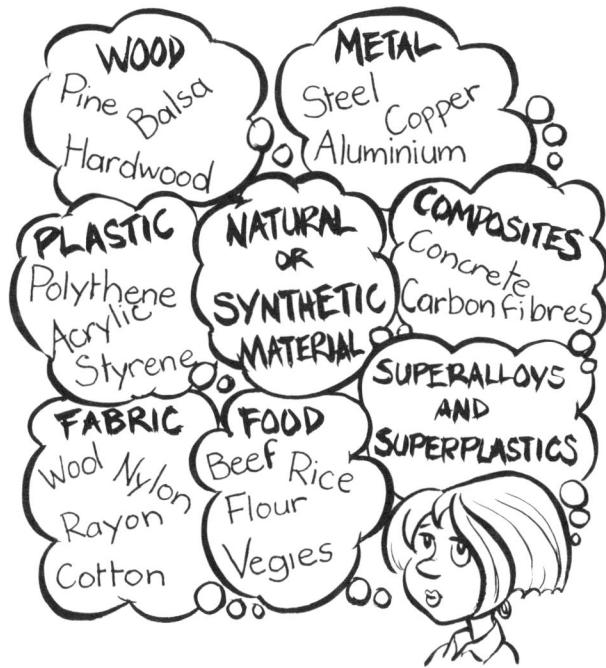

Fig. 6.2 There is a wide variety of materials from which to select.

FACTORS AFFECTING SELECTION

The five most important factors to consider when selecting your materials are:

- cost
- physical properties
- availability
- methods of manufacturing
- compatibility with other materials to be used.

Here are some questions you may need to ask when selecting, testing and evaluating materials:

- What are the main properties of the material?
- Is the material readily available?
- What size and form of material is required?
- Will the product be used indoors or outside?
- Will the product require a protective finish?
- Will it need strength to hold a load?
- Does it need to be light?
- What is the cost of the material?
- Is the material easy to shape?
- What joining methods are required?
- Who will be using the product?
- Is the material compatible with other materials?

These and many other questions need to be raised continually as you progress through your project.

COST

Throughout your technology studies, and particularly for your own production work, you will be required to produce lists detailing the cost of materials . The task of putting together a costing list is not difficult. However, you need to be very careful with measurements and quantities, and to make sure you use up-to-date and accurate materials price lists.

The MATERIALS COSTING worksheet on page 82 is suitable for costing wood, metal and plastics. Other versions of this sheet may be required for other materials or accessories.

PROPERTIES OF MATERIALS

When making decisions on the materials best suited for your particular project it is essential for you to find out about the properties of the material. If this is not done, you could quite easily use a material which is inappropriate for the particular job. Information can be gained through discussions with your teacher, through your own past experiences and observations, by reading technical books and various craft magazines and, of course, by *testing* the materials for yourself.

As you learned in Chapter 5, the properties of a material are its distinctive abilities or qualities. Properties determine the means of use and the overall makeup of the specific material. Many materials have a number of different properties, while others may have just one or two. The list below does not contain all of the possible properties. Many others can be found in other technical publications.

- hardness
- mass
- toughness
- appearance
- compression
- shear strength
- tensile strength
- malleability
- atmospheric resistance
- conductivity
- brittleness
- ductility

If you are unsure of the meanings of these properties, ask your teacher or look them up in the Mini-dictionary at the back of this book, or another dictionary, or other technical books.

COMMON MATERIALS

Materials are available in a variety of sizes, shapes and forms. Listed below are some of the more common materials which can be purchased at your local hardware store or materials distributor. In most cases, the materials you will be using will be available through your school.

WOOD

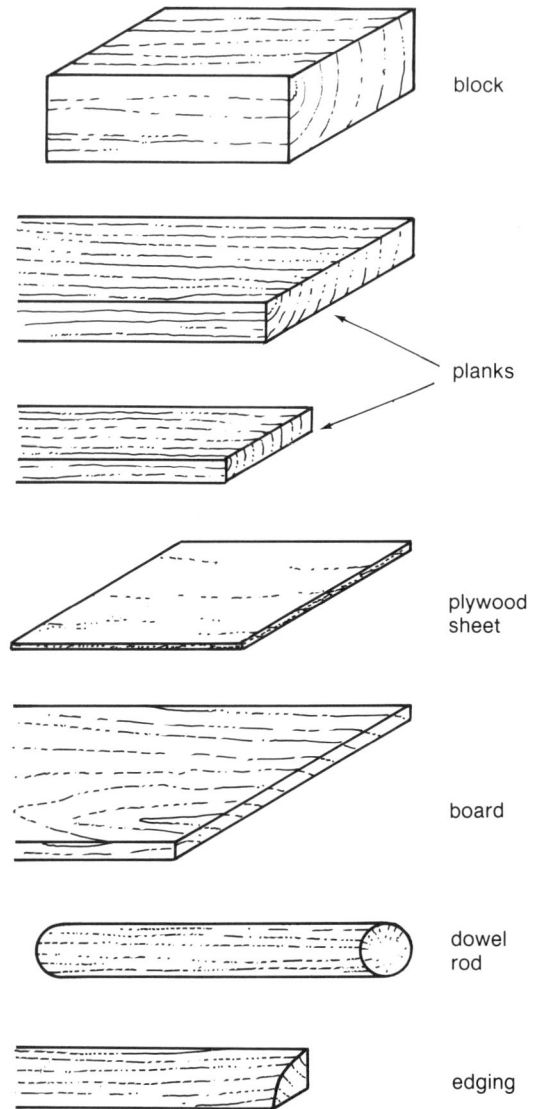

block

planks

plywood sheet

board

dowel rod

edging

Fig. 6.3

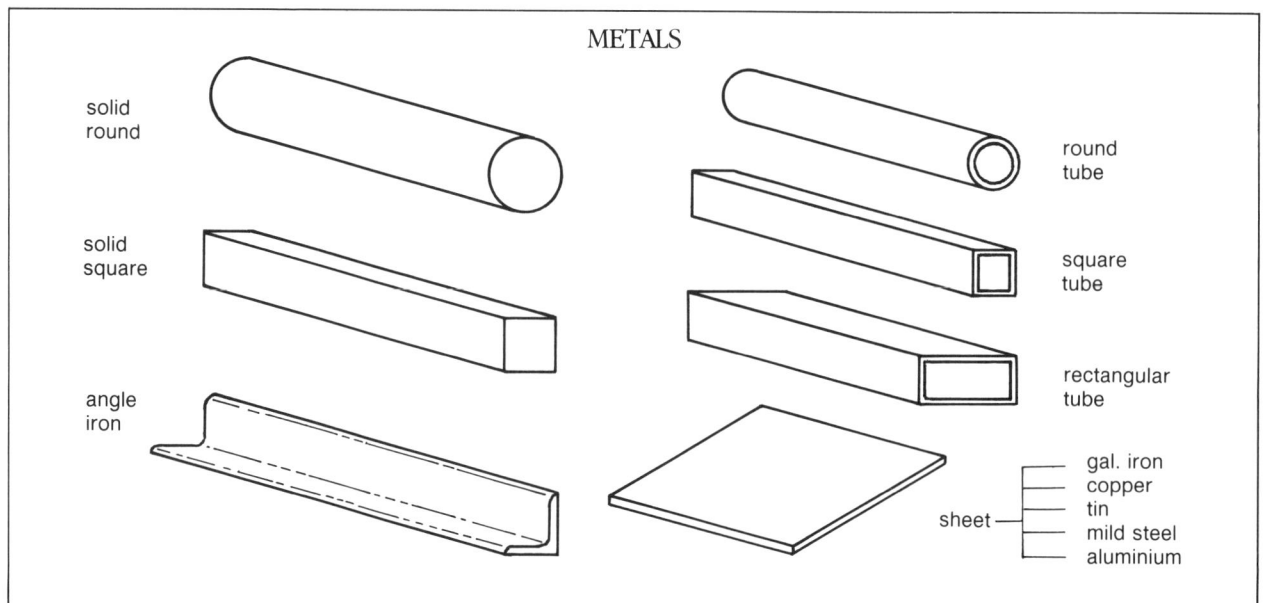

METALS

solid round

solid square

angle iron

round tube

square tube

rectangular tube

sheet — gal. iron
copper
tin
mild steel
aluminium

Fig. 6.4

PLACTICS

FIBRES AND FABRICS

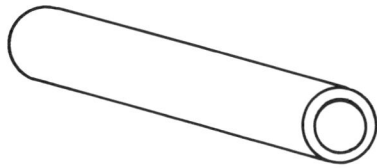

round
tube

liquid
resin

crushed
granules

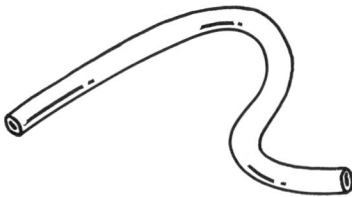

hose

solid
tube

solid
square

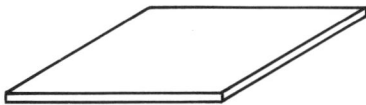

sheet

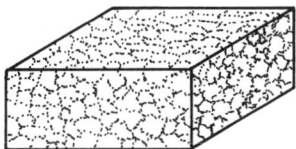

styrene

recycled

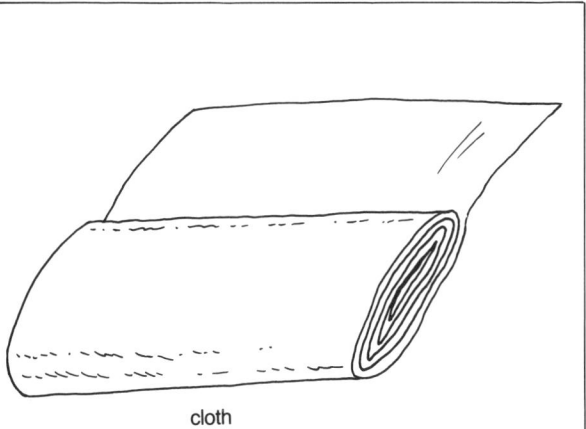

cloth

thread

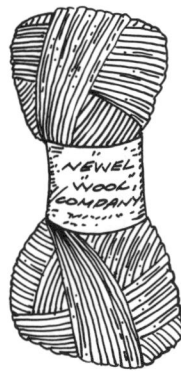

wool

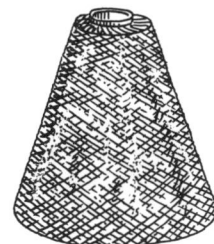

yarn

Trims

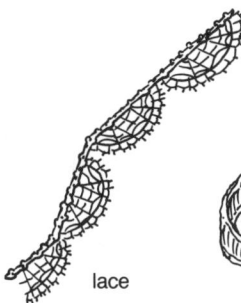

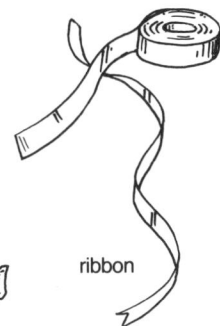

lace

elastic

ribbon

Fig. 6.5

Fig. 6.6

OTHER MATERIALS

There are a number of other materials you may wish to use. However, it is impossible to have all of these available in the classroom at all times. Many of these will be used only as secondary materials which assist in the overall function or appearance of your project, for example: concrete, plaster, cement sheeting, ceramics, leather, enamels, glass.

MATERIALS TESTING AND EVALUATION

Whenever you buy a product you make an assessment on whether or not the product is suitable for the specific need you have. Sometimes this is made easier for you by the advertisements for the product. However, these can be misleading and much thought needs to be put into your decision to buy.

Many products only partly carry out the required task or are unable to function over a long period of time without breaking or breaking down. You then have to buy replacement parts, have the product repaired or buy another one. The poor functioning of the product is often caused by the use of inferior materials.

Most companies that are involved in making products for consumers spend hundreds of hours and thousands of dollars developing and testing the function and materials of their products. These are the aspects which can often determine the success (viability and profitability) of the product in the marketplace. When making your own products you, too, will need to take steps to ensure that correct materials are used.

Experience is one way of knowing whether a material is suitable or not. You may also find the information in magazines or books, or you may actually test the material to see if it is capable of carrying out the task required. Testing of materials should be a part of your classroom work as it will enable you to learn about this part of the selection process and it will also assist you in understanding the properties of specific materials. It is no good finding out *after* your project is started that you have used the wrong material. It is worse to find out after you have completed it! It makes much more sense to test and establish whether or not the material will suit your needs and if you are able to form and construct the product using the selected material.

It is a good idea to conduct a number of tests on a variety of materials. Your teacher will probably help you out with at least one of these tests. In doing so, he or she will set ground rules or guidelines to indicate what is expected of you when conducting your own tests. The tests could involve various types of materials: some common, others less common; some similar to each other (comparative tests); some unusual or relatively unknown materials. Tests may be carried out on the particular material(s) you propose to use in your own project. Tests will assist you not only to learn about materials, but will improve your ability to use tools and equipment.

Most tests that you can conduct in the classroom are fairly simple. For example, you may test a particular material (e.g. pine) for weatherability if it is to be used in the manufacturing of outdoor furniture. After putting it through both a wet and dry test — and with and without protective coatings — you can assess the viability of the material. You may, perhaps, test and assess the capability of various fabrics to be machine sewn, or their shear strength (their tendency to tear under stress). You can then make judgements from these assessments — that is, evaluate the material.

Here are some questions to ask when testing and evaluating a material:

- Why is it being tested?
- How is it being tested?
- What are the expected outcomes of the tests?
- Were the expectations fulfilled? If not, why not?
- What problems surfaced during the test and how were they overcome?

SAMPLE TESTS

The type and the number of tests you do will be up to you and your teacher and will depend on the requirements of your course or the project being attempted. Tests might be done to check the following:

- weatherability of timber
- corrosion of metal
- suitability of adhesives to be used on various materials
- suitability of finishes such as paint, lacquer and oils on various materials
- hardness
- shear strength
- tensile strength
- structural strength
- conductivity (heat and electricity)
- toughness
- elasticity
- compressive strength
- flammability of fibres and fabrics
- stretch (fabrics)
- brittleness

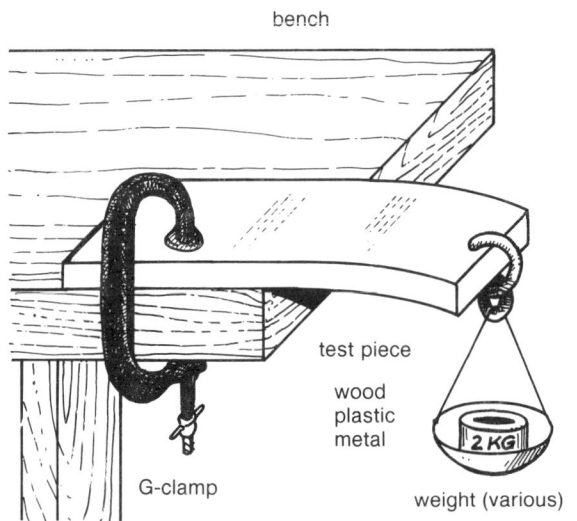

Fig. 6.7 Stiffness. Each material should be placed under a variety of weights. The material which has the least deflection is the stiffest.

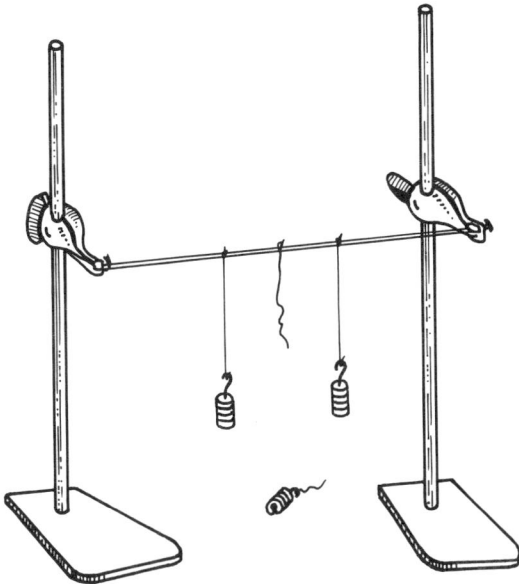

Fig. 6.8 The tensile strength of various fibres can be tested and measured by adding weights until the thread breaks.

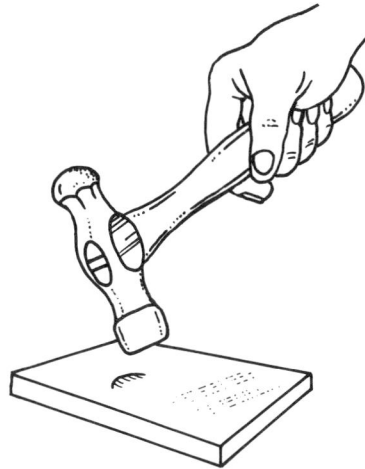

DRILLING

stainless steel

copper

steel

SAWING

tube

bar

angle

ATMOSPHERIC TEST

3 different metals, dampened and left outside for 3 days. Record daily progress.

Fig. 6.9 There are many simple tests which can be carried out on metals.

metal scriber

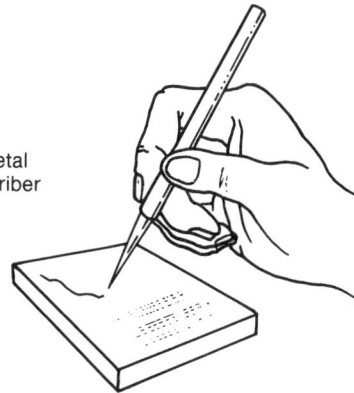

Fig. 6.10 Hardness. Blades need to be hard to enable them to cut easily and stay sharp. Items which are difficult to scratch or dent are also hard.

RECORDING YOUR FINDINGS

You will find that keeping records of tests and comparative tests is much easier and more readily understood if a standard format is followed. You might like to use the MATERIAL TEST sheet provided on page 79. Note that this sheet includes space for a written evaluation on the tested material.

REVISION QUESTIONS

1   Explain what is meant when we refer to a 'natural material'.

2   Name *three* natural materials.

3   What is an alloy? Name *two* alloys.

4   What are the *five* factors to consider in material selection?

5   Explain what is meant by the word 'property' when applied to material. Give *one* example.

6   Circle TRUE or FALSE:

   a   The first materials used by people were natural ones.   T   F

   b   Plastic is a natural material.   T   F

   c   Copper is a composite material.   T   F

   d   Alloy means aluminium.   T   F

   e   Hardness may be tested by scratching the surface of a material.   T   F

   f   Steel is found in the ground.   T   F

7   Explain why it is necessary to test materials before using them in making products.

8   Give *two* examples of tests which could be conducted to test the shear strength of (a) a fabric and (b) a metal. Name the metal and fabric you select.

STUDENT ACTIVITY

The ten items shown below are made from materials which have distinct properties. It is these properties which determined the selection of the particular materials for each item.

Name the material(s) that are used in each item and list *two* main properties of *each* of the materials. Say how these properties assist in the function of the item.

School bag

Garden hose

Stackhat

Skateboard

Jug

Food wrap

Drink can

Kettle

Kitchen tap

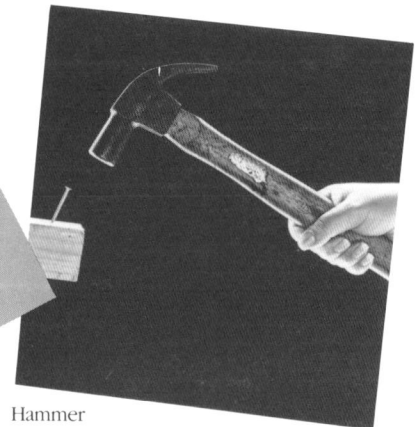

Hammer

7

PRODUCTION AND PLANNING

When you embark on your production, the processes you use to make your product will depend mainly on the material to be used and the purpose of the product. For example, when you are working with wood, you will use certain joining methods. These methods may be quite unsuitable for joining plastics or cotton fabric. It is important that you choose the best method for the job in hand, and to do this you will need information. You may have knowledge from your own previous experiences, or you can acquire it from classroom demonstrations, by asking people, or reading craft and technical journals.

Planning is also an important aspect which needs effort and thought. It is much easier to make things if you have a clear plan of attack. This will assist in resolving any problems which may surface at this stage.

This chapter will focus on:

- the need to plan and organise your time while working through your project
- the preparation and planning required before and during the making of your product
- the skills, techniques and knowledge needed in the making of your product
- processes used in the making of your product
- the importance of evaluating your product
- safety awareness when using tools and equipment.

PLANNING

It is essential to plan your project carefully and thoroughly. Often, mistakes can be avoided by thoughtful planning. A plan will also help you to organise your time and it will enable you to keep a continual check as you work through your project.

First, write down a *broad plan*. If you will require any special materials or parts, make a note so you will be sure that they will be available by the time you need them. In your planning, check that all drawings you are using are accurate and contain enough detail to enable you to plan ahead.

Once you have made your broad plan and begun work on your project, it is a good idea to keep a record of any changes you make — and the reasons for the alterations. Do this on a regular basis as you progress through your project. This information will be useful at a later date.

A PRODUCTION PLANNING sheet is provided on page 80. The MATERIALS COSTING sheet on page 82 would also be useful for projects involving wood, metal and plastics. (There is more on costing on page 30.)

The broad plan is fine for long-term planning. However, it will also be necessary for you to plan your work on a *short-term* basis — probably weekly. List what you think you can achieve in the time available, and write down any tools and equipment which you may need to assist you in your work. It is important to record this (in your journal/logbook) as it gives you a plan to follow and also a goal to work towards. (See the JOURNAL / LOGBOOK ENTRY sheet on page 71.) Your working time can then be used more efficiently.

Here are some questions you may need to ask yourself before and during the production stage:

- What is the specific purpose of the product?
- Where will it be used (indoors/outdoors)? Is weathering a factor to consider?
- Is it required to match another item?

- Do safety aspects need to be considered?
- Who will be using the item?
- What size will the user be: big? small? skinny? fat? just average? (ergonomics)
- Are aesthetics important? Does the product have to be beautiful?
- What processes are required to form the material?
- Will the item be under any particular stress (e.g. weight, water pressure, or air pressure)?
- How will it be put together?
- What tools will I need?
- What equipment will I need?
- Is the material I need available?
- Do I have the skills to make this project?
- Where can I obtain information which can assist me with this project?

Fig. 7.1

PRODUCTION

When you come to the making of your own products, all of the learning you have done since the initial planning stage comes into use.

If you have completed all of the relevant learning through activities such as design exercises, material testing and selection, research and investigation work, as well as any other requirements set down, you will have little trouble in completing your project to an acceptable standard. However, before you rush into your production work, it is also important that you acquire some basic skills in the use of the tools, equipment and processes required to make your project.

Fig. 7.2

EQUIPMENT, SKILLS AND PROCESSES

You will probably have had some experience with tools and various pieces of equipment. If not, or you are unsure of your skills, your teacher will give you demonstrations and small practice exercises which will enable you to learn or upgrade your skills. As well, you will be taught about any particular techniques and processes which will be of use in your project.

On various occasions, your teacher may ask you to find relevant information about the skills and processes that you require for your project. If so, you will need to consult relevant books and other texts (hobby books, trade manuals, technical magazines, etc.) as well as worksheets or skills booklets compiled by your teacher.

You may learn some of the skills and processes before you commence your project, but most of them you will learn as you work through it. The best method of acquiring skills is to practise them on smaller tasks first or to use sample pieces to try things out.

Fig. 7.3

The types of tools and equipment used in the following and other processes (this is not an exhaustive list) will vary, depending on the material being used and its properties.

MARKING OUT	CUTTING	SHAPING	FORMING	JOINING	FINISHING
measuring	drilling	moulding	bending	welding	painting
scribing	sawing	forging	folding	bonding	polishing
squaring	turning	casting	pressing	sewing	staining
templates	shearing	knitting	twisting	seaming	enamelling
patterns	chiselling	chiselling	vacuum	nailing	lacquering
	filing	filing	forming	screwing	coating
	tapping	planing	raising	rivetting	sanding
	milling	gouging	hollowing	bolting	hardening
	grinding	grinding		dowelling	
	routing	routing		soldering	
	planing	hammering			
		pressing			

MARKING OUT

Accuracy is the most important aspect when marking out. Tools such as squares, rules, gauges, tapes and dividers are just a few which will assist you when marking out.

Remember, if you are not accurate with your measurements, your product will not be completed to a very high standard. The examples below should assist you to achieve the accuracy you require.

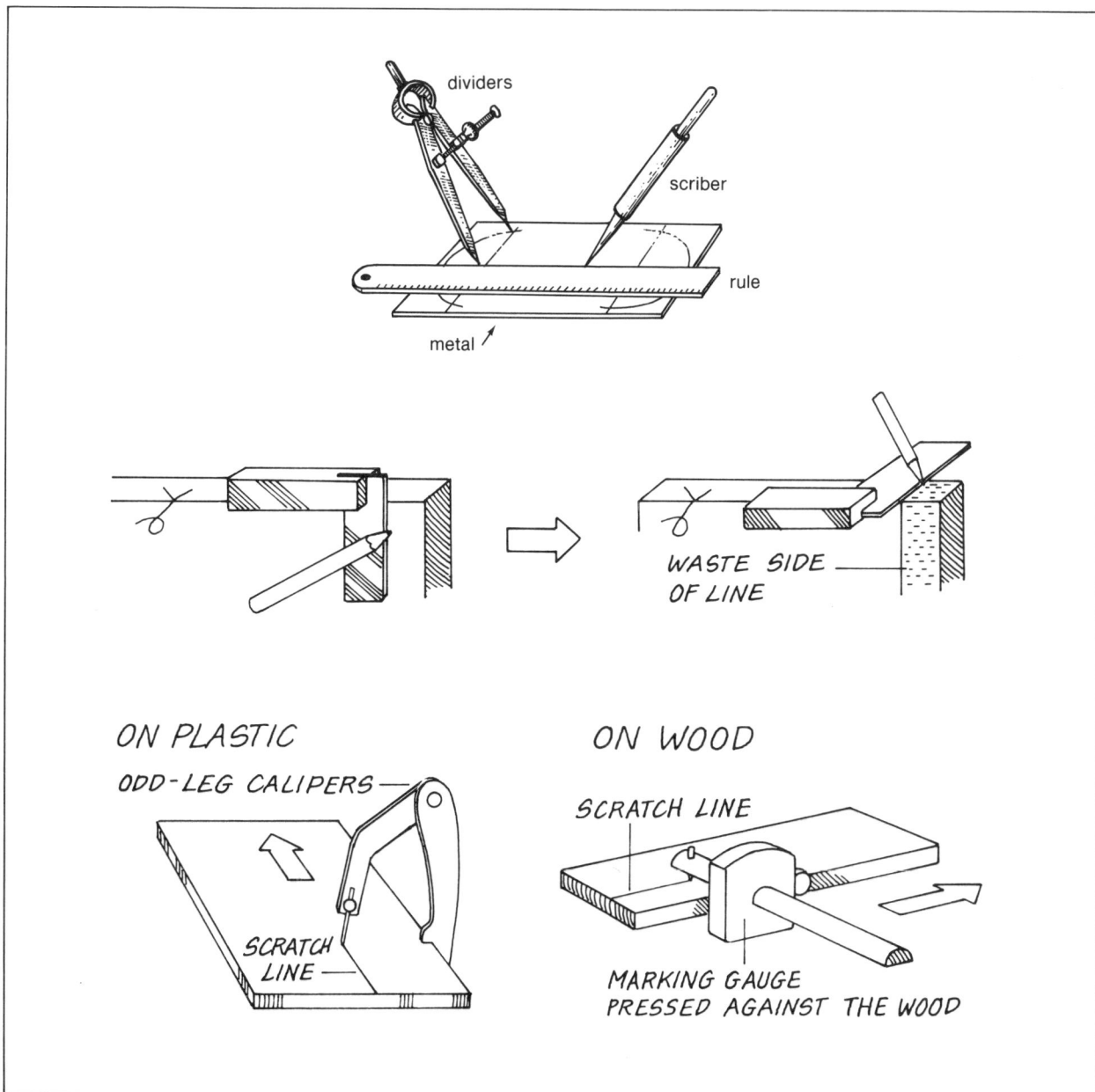

Fig. 7.4 Marking out

CUTTING

Cutting can be done by using a number of different tools and equipment such as saws, drills, knives, scissors, tin snips, punches, chisels, guillotines, planes, grinders and sanders. When using this equipment, always check that it is in good condition. Blunt tools cause poor quality cuts and make the task difficult. Shown below are a few examples of cutting tools used on various materials.

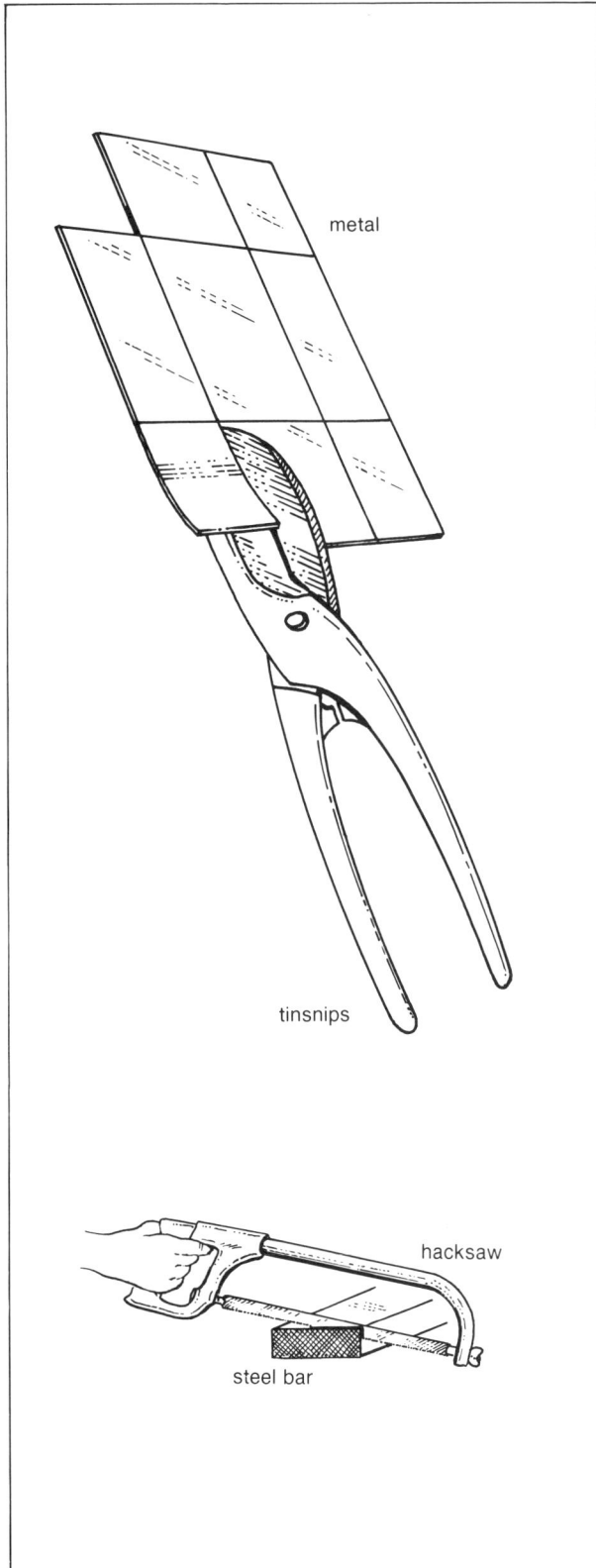

Fig. 7.5   Cutting

SHAPING

Shaping is generally required when you wish to achieve a curved, round or odd type of shape. The manner in which you are able to achieve this will again depend on the type of material you wish to shape, and also the availability of equipment. Shaping can be achieved by using simple tools such as a hammer or mallet (for example, to hollow or raise a metal dish) or with more complicated equipment such as a press which can press out car panels or hard plastic covers for products (like television sets and computers) which require durable protective covers.

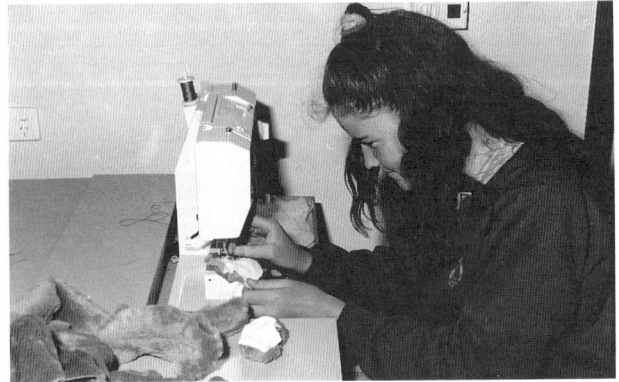

PLUG AND RING MOULD

The plug should be slightly tapered. The diameter of the ring must allow for a thickness of plastic all the way round. The plastic is softened and then the ring presses it over the plug to form a plastic dome.

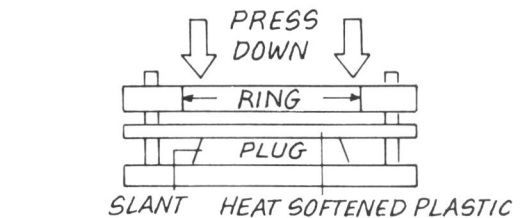

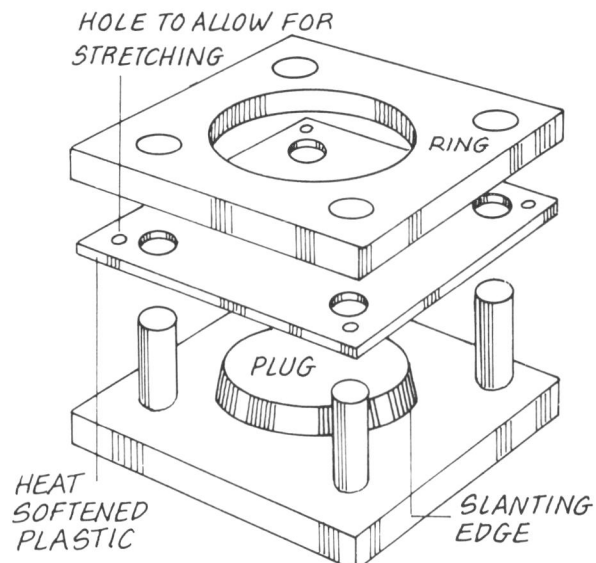

Fig. 7.6   Shaping

FORMING

Forming of materials is achieved through folding, bending, twisting, raising and pressing. Forming is necessary to alter one form of a material (e.g. a flat metal sheet) into another form (e.g. a box).

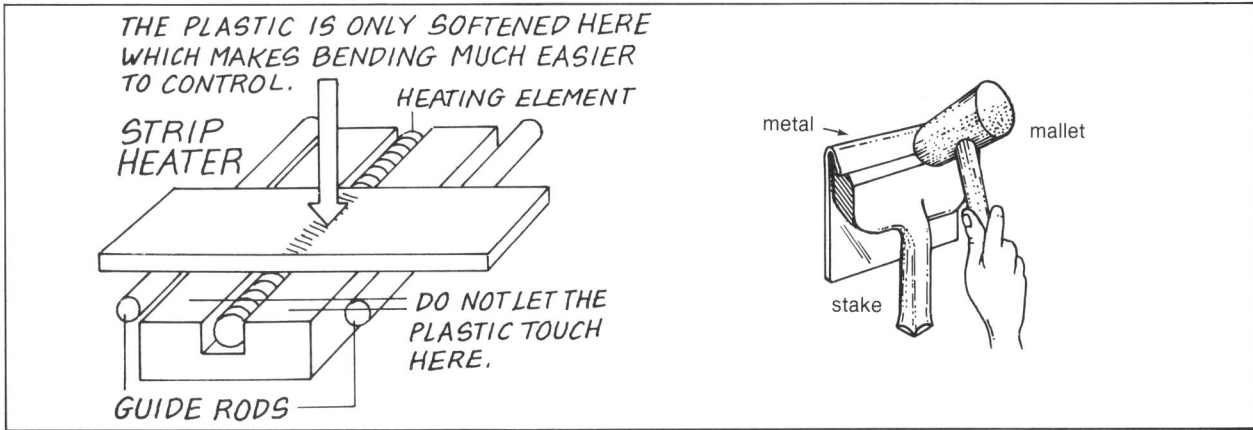

Fig. 7.7 Forming

JOINING

Your joining methods may well determine how successful you have been in fulfilling the initial project brief. It is essential you take care when selecting the method so you can complete this part of your project to a high standard. The variation in methods used for joining metals, plastics, fabrics and wood are enormous. This is even more so *within* a particular material group. For example, in joining mild steel, the following methods can be used: rivetting; self tapping; welding; bolting; seaming.

Take care when deciding on which joining method is most appropriate for your particular project.

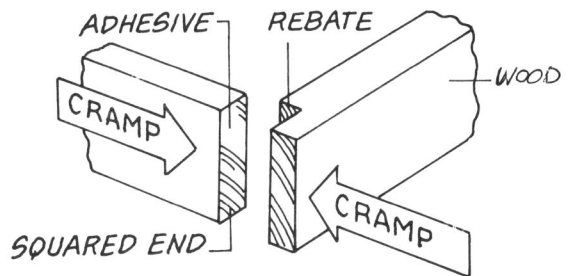

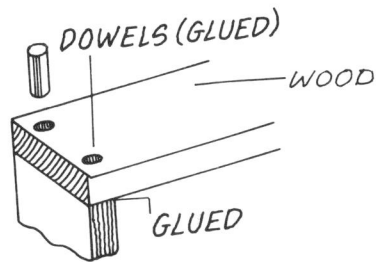

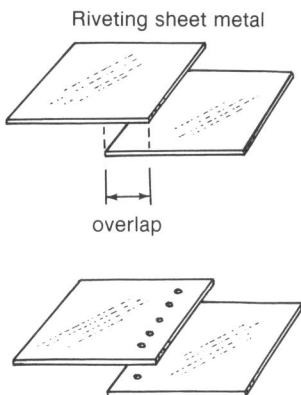

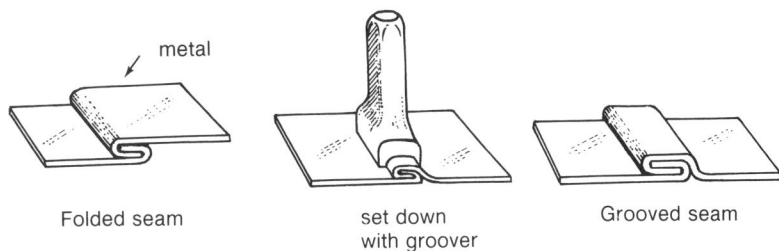

Fig. 7.8 Joining

FINISHING

The finishing of your product is important for the final presentation of your project. The visual appearance of many products is a very important aspect, and consideration of finishing methods should not be taken lightly. In many cases, the type of finish will be determined by the use of the product where and how it will be used. For example, products to be used outdoors will require a finish which will make them weatherproof. Metals such as mild steel need to be coated so they will not rust; pine requires coating with a weatherproof paint or lacquer to prevent rotting and warping.

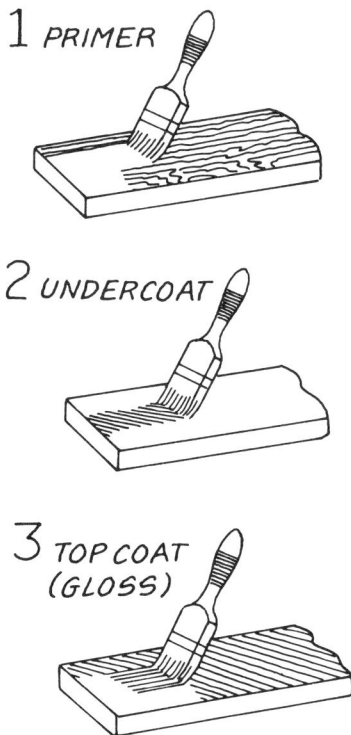

1 PRIMER

2 UNDERCOAT

3 TOP COAT (GLOSS)

Fig. 7.9   Finishing

SAFETY

Working in workshop areas is generally a very enjoyable experience which finishes in you making a product which you are able to keep and take home. But to be enjoyable it has to be safe — for everybody.

Many of you will have learnt certain safety rules and precautions while working in earlier classes. If so, you will only need to revise these and learn any new rules which apply to the use of equipment that you have not previously handled. Tools and equipment are only dangerous if they are not kept in good working order or are being used incorrectly. Always report any equipment that may become damaged while you are using it. Your teacher would prefer you to be honest than for someone to be hurt because you put faulty equipment back in the storage cupboard.

If you use your common sense, the workshop will continue to be the enjoyable place for learning it has always been. Here are a few basic safety rules:

* Wear protective clothing at all times when working in the workshop.

* Tie back long hair and remove any loose clothing. It is easy for these to catch in machines.

* Never run around or fight in the workshop.

* Always ask for assistance if you are unsure of how to operate equipment correctly.

* Wear protective glasses or shields when operating drills, grinders, linishers and other powered cutting equipment.

* Make sure all machines have the necessary safety guards.

* Know where emergency stop buttons are located.

* Always clamp down materials that are small or are difficult to hold when drilling and so on.

* If you have an accident, make sure you inform the teacher. Know where to go for First Aid. Know some basic First Aid.

Fig. 7.10

SURE AND STEADY

Here are some further points and reminders to consider in your production work.

- Make sure you have a work plan for each session. This will give you an indication of the time, materials and equipment you will be needing.
- Set yourself realistic goals. If you attempt to achieve too much in one session you will inevitably make mistakes which will set you back.
- The faster you work, the more chance for mistakes. A good steady pace is sufficient and allows you time to think about each step in the process.
- Don't take short cuts.
- Make sure you work accurately. If you don't, you reduce your chances of successful completion.
- Assist others in the class when required. You never know when you may need a similar favour at a later date.

Fig. 7.11 If in doubt about your construction methods or sequence, ask your teacher.

PRODUCT EVALUATION

As you work through your project you will need to evaluate it on a regular basis. Doing it regularly will prevent you from getting to the completion stage, finding that the product does not meet the initial intentions, and having to start all over again! In the case of units of work which have time limitations, you may not be able to complete your work in time for assessment.

On the completion of your project you will need to evaluate the product against the original brief which was drawn up. This is done to give you a clear picture of the way you went about producing your project and will give you information which may be used for a similar project at a later date. Check through your folio notes and the production recordings in your journal/logbook to compare the initial plans and alterations against the completed project.

A PRODUCT EVALUATION sheet is provided on page 81. Use this to help you record your findings.

Many of you will continue with other technology studies. Hopefully, you will have learnt from your experiences, and any problems which you encountered will be easier to solve next time. The evaluation of your product will also help you in your future decision-making on the selection of different tools, methods of construction or the finishing applications for the next project.

REVISION QUESTIONS

1 Give *two* reasons for the need to plan your production work.

2 Explain what a process is.

3 Circle TRUE or FALSE:

   a When using machinery, you should always be aware of the location of the emergency switch. T F

   b It is always a good idea to practise your skills before commencing your production work. T F

   c If you are unsure about how to operate a machine, ask your friends to show you. T F

4 List *six* safety rules or guidelines for working in the workshop.

5 Complete this sentence: 'It is important to evaluate my project because…'.

STUDENT ACTIVITIES

Throughout the making of your product you will have used a number of different tools and processes. These will have made your task much easier and simpler.

1 Name *two* of these processes and write a short summary (approx. 30 words) on how each process assisted in making your task easier.

2 Choose *two* of the tools you used in your production work. Explain how each one helped to make your work easier.

8

MODEL MAKING

Designers and other people involved in making products sometimes use more than drawings to show people how a product may look. They also use models — prototypes and mock-ups. Such models can also be helpful in showing up possible problems before too much work has been put into the actual product.

A prototype is a realistic working model and is generally used to assess whether or not the product will be able to carry out its intended function and use. A mock-up is used in a similar way but it is generally used to give an indication of how the product may look, the way it may be joined, and so on. The main difference between the prototype and the mock-up model is that the prototype is more realistic and closely resembles the finished product.

This chapter will focus on:

- the use of prototypes and mock-ups in the making of your own products

- how to go about making prototypes and mock-ups

- materials, tools and equipment required to make prototypes and mock-ups.

During your own production work, you may find it useful to make models of your product. They can be made from various craft and modelling materials.

Making mock-ups and prototypes of products is very useful for a number of reasons. Once the product is 'off the drawing board' and in a 3-dimensional form, you can see what it will look like — you can visualise the actual finished object. You are then able to test various aspects of the design and make decisions on changes to that design or, perhaps, to the proposed construction.

For example, you could try different ideas for the support rails on your table, or different styles for the bag you are designing. Or a colour clash may show up when the model is painted, or the mechanism doesn't operate the way you expected it to. If you are making an item which requires complex folding procedures (for an example, see page 26), you can use a paper mock-up to practise on first. This will allow you to understand the folding procedures without wasting valuable materials. All of these problems and many more can be solved through producing mock-ups or prototypes.

If the product is large, the model can be scaled down — made smaller. If it is very small, the model may need to be scaled up (an enlarged version) to assist in checking its function properly. The scaling of the model could be 1:1 (the model is the same size as the finished item); it could be 1:100 (the model is one-hundredth of the size of the finished item); or it could be 10:1 (the model is ten times larger than the finished item).

Fig. 8.1   Mock-up models on a 1:10 scale

MOCK-UPS

The making of a mock-up should be kept in perspective in terms of the product being made. If you are making a simple item, the mock-up does not need to be too complicated. If you are making a product which has many parts or is complex in its construction, then it may be necessary to make up a fairly detailed mock-up. You will need to assess this at an early stage.

PROTOTYPES

When making a prototype you will need to be more accurate with the construction, complete it to a functional standard and in a realistic manner. A prototype can be full-size or scaled up or down, and is generally the first working model of a product. Manufacturers of motor cars use prototypes of their vehicles to test them for function, safety and so on, as well as using them for marketing purposes.

Fig. 8.2   Producing a prototype of a Holden car

MATERIALS AND EQUIPMENT

If your product is to be made from metal or plastic, you could use paper or cardboard of varying thickness. This could be coloured to simulate the type of metal or the different coloured plastic to be used. To simulate different types of structural steels, you might try soft modelling wire or straws. Lengths of thin wood, such as matches (with the heads cut off), may be painted black to look like square tubular steel. These and many other materials will assist you in making your mock-ups and prototypes look realistic.

The making of these models is very similar to putting together cars, aeroplanes and various other items from modelling kits. If you have had experiences with this type of work will know how enjoyable and rewarding it can be. You will be able to think of many other methods and materials which can be helpful in this stage of your production work.

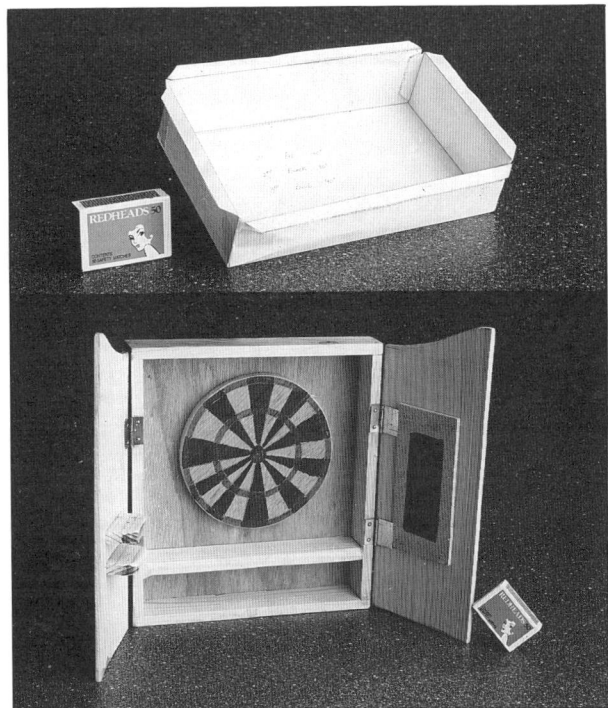

Fig. 8.3   Making mock-ups is similar to making models from kits.

At the end of this chapter you will find some ideas that may help in assisting you with this task. Listed below are some materials and equipment that are easily obtained and useful in your model making.

cardboard	paper	clay/plasticine
drawing pins	rubber bands	scrap metal
aluminium cans	plastic bottles	Lego Technics
polystyrene cups	Meccano sets	glues (various)
clips (various)	sticky tape	string/cotton
masking tape	Textas	pens/pencils
plastic tubing	thin plastic	sewing pins
modelling wire	plaster	sponge
drinking straws	paint	wood offcuts
vinyl	balsa wood	shoe boxes
pipe cleaners	icy-pole sticks	milk cartons
Stanley knife	drills (various)	coping saw
cutting board	small clamps	plyers
matchsticks		
(with heads cut off)		

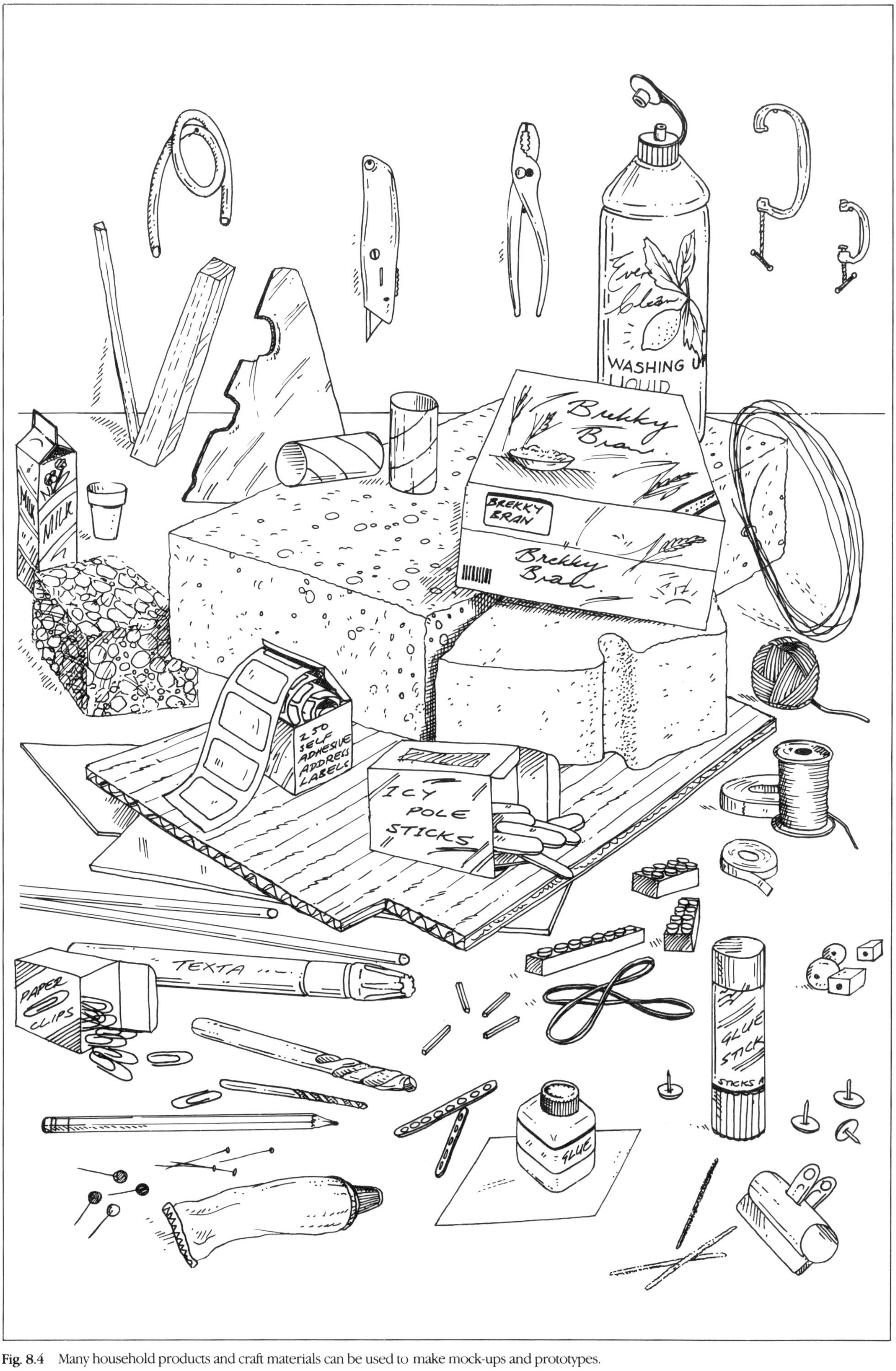

Fig. 8.4   Many household products and craft materials can be used to make mock-ups and prototypes.

REVISION QUESTIONS

1 Explain the difference between a prototype and mock-up.

2 What are the reasons for making prototypes and mock-ups of a product?

3 Name *one* industry that uses prototypes before making their products.

4 Circle TRUE or FALSE:

   a Mock-ups can be used to give potential clients an indication of how a product functions and looks.   T   F

   b Prototypes are used only in the motor car industry.   T   F

   c A prototype is often the first working model of a product.   T   F

   d Designers use mock-ups to show clients what products may look like.   T   F

5 Explain what is meant, in model making, by the term 'a scale of 1:20'.

STUDENT ACTIVITIES

Listed below are five exercises which will assist you in learning about making models. Select *one or more* to complete.

In two of these exercises the models are to be reduced to a specific scale — 1:10. This means the model will be ten times smaller than the finished product. All the *parts* for the model, therefore, will need to be scaled down, too. If you have any problems with this, see your teacher.

1 Use paper or cardboard to make a 1:10 scale model of a metal box. The finished metal box is to be 400 mm wide, 1000 mm long and 500 mm high, and has tabs or folds to assist in joining it together.

2 Design and draw up a video cassette rack to be made from wood or plastic (acrylic). The rack is to hold 10 cassettes when completed. Once you are happy with the design drawing, make a 1:10 scale mock-up of the rack. Remember, you will need to measure the size of a video before you commence.

3 Use clay to design and make a model of the body of a new type of two-door super-sleek sports car. You need only concern yourself with the exterior appearance — not the mechanics or function.

4 Make a prototype of a fold-away chair. Use the sketch in Fig. 4.2 to assist you in this task.

5 You may already have drawn up a product design of your own, or your teacher may have provided one for you. From the information on your design or drawing sheet, make up a mock-up or prototype (whichever is most appropriate) of the product.

CHAPTER 9

RESEARCH AND INVESTIGATIONS

Research and investigations can help you to obtain information which you may need for the making of your product and to help you compile reports on other issues related to your technology studies.

In this chapter you will be given information about methods, formats, and planning to assist you in formulating your assignments, recording your findings and presenting your reports.

This chapter will focus on:

- the relevance of research and report writing in technology studies
- the value of planning when doing investigation work
- the methods and formats used in producing and presenting investigation reports
- how to locate, gather, analyse and evaluate information and data.

Research and investigation work is an important aspect of your technology education. In your own work, as in industry, it is often necessary to gather information and data before you can come up with solutions to problems.

There are numerous ways you can go about obtaining this information and many sources of it. This chapter cannot deal with all the relevant methods or sources that you may use but, after reading it, you should gain sufficient knowledge to enable you to tackle the research and investigation for your own needs with confidence.

THE PURPOSE OF INVESTIGATIONS

In your technology classes you will probably be involved in carrying out research or investigations for two purposes:

- finding information directly required in your own production work
- investigating other issues to add to your general knowledge of the world of technology, and provide information which you may need in your future production work.

The end product (apart from the knowledge you gain) of these investigations will probably be a *report* of your findings. The report may be part of your assessment or grading, so it is essential that you tackle the task with care.

INVESTIGATING FOR YOUR PROJECT

Throughout your production work, you will constantly run into problems and need more information to assist you in finding solutions before continuing. For example, you may need to know which material is best to use, or what is the best method for joining the parts of the product. Whatever the problem, you will need to go about finding the information in an organised and systematic way.

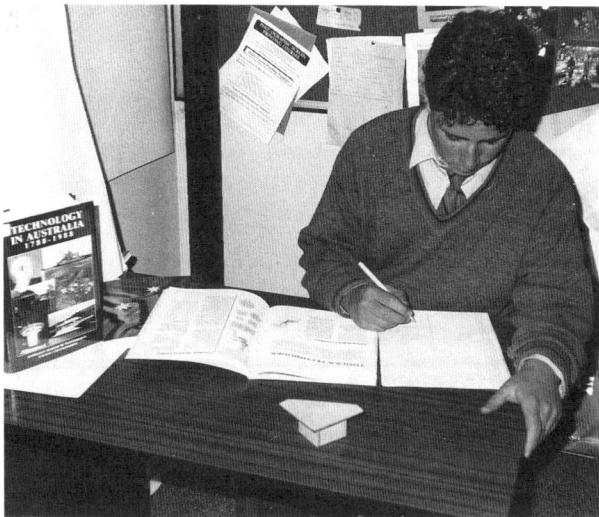

Fig. 9.1 A student working on research for his own product. Note the reference book and mock-up.

INVESTIGATING OTHER ISSUES

You may be required to investigate and write reports on various technological topics which are not directly related to your own production.

These assignments will usually be given to you by your teacher with, perhaps, some input from you. They are designed to give you practice in research skills and to help you gain knowledge of aspects of technology beyond your immediate needs. For example, you may be asked to research and report on new or unusual materials and how they were developed, or the feasibility of recycling plastics, or the impact of the transistor on people's lives.

TOPICS FOR INVESTIGATION

Under the three areas listed below are examples of topics which you may need (or be required) to investigate.

PRODUCTION – materials; tools and equipment; methods of fabrication and manufacture; cost factors; safety.

DESIGN – ergonomics; appearance; function; marketing; product analysis; new materials; safety; consumer appeal.

OTHER ISSUES – the impact of technology on the environment, society and the economy; new materials; inventions and innovations; links between science and technology; health and safety.

These lists are not exhaustive. Many other topics that relate either directly or indirectly to your own projects could be researched and investigated. Remember, if you are given the chance to select a topic, it is useful not only to select one which is relevant to your own project, but also one which is of interest to you.

KEEPING RECORDS

You should always keep a record of where information was found, the date, and the name of the source (person, book, etc.). The INFORMATION RECORD sheet supplied on page 82 should be filled in at the time you collect the information — don't rely on your memory! The sheet should be kept in your folio for future reference.

INFORMATION RECORD

TOPIC .... *Properties of materials* .... DATE *20/3/92*

Source:

Smith, K.P., Materials and Design, Daily Publishing Company, Australia, 1990 (Secondary source)

Chapter/Volume: *Chapter 6*   Pages: *56–78*
(if applicable) *(materials and properties)*

Type of information:

Description of various materials and their uses.

Fig. 9.2

PLANNING YOUR INVESTIGATION

It is essential for you to have a sound understanding of what is required of you, and to prepare and plan ahead if you are to complete assignments and reports *on time* and to a *satisfactory standard*. Students who find it difficult to write up reports will find that this is generally due to a lack of planning and poor researching of subject matter.

An organisational plan is essential. One such plan is outlined below. Your teacher may prefer you to use another framework but the steps will be similar.

1 Define the aims of your investigation.

2 Plan your approach to the investigation.
   (If possible, decide on the form your report will take.)

3 Locate information and data.

4 Gather and record information and data.

5 Analyse the information and data.

6 Evaluate the information and data.

7 Draw conclusions.

8 Prepare your report and summary sheet.

AIMS AND APPROACH

Your investigations must be accurate and precise. Before starting, make sure you are quite clear on *what* you wish to find out and *how* you will proceed.

The subject of the investigation (the topic) may determine the kind of presentation you choose for your report. Details of the various ways you can present your report are on pages 57, 58 and 62. Be familiar with these *before* you start your investigation.

Your investigation work may be undertaken individually or, if permitted, as a group activity. A workable number for a group would be no more than three. If the report is to be assessed, each member of the group should submit an individual report or piece of work so the teacher can identify and assess each student's contribution.

The questions below may assist you to make some early decisions. If you have other concerns write them down and then work through them one at a time. Doing this can help sort out many problems before they occur.

● What do I wish to find out during this investigation?

● Where do I find the information I am looking for?

● What is the difference between a primary and a secondary source?

● How do I go about borrowing books or other equipment and materials?

● Who is the best person to ask about this particular issue?

● How much work does my teacher expect of me?

● What format could I use to present my findings? Is this set by my teacher or can I choose my own form of report?

● What (if any) is the due date for this assignment?

LOCATING INFORMATION

There are many ways of finding the information you require but no matter which way you choose, it is important that you are well organised and follow a procedure in the search.

There are two kinds of information sources — primary sources and secondary sources — and each requires a different approach and organisation to use.

PRIMARY – A primary source is one that provides *unprocessed* (first-hand) information — that is, information that you have gained from personal experience or observations.

SECONDARY – A secondary source is one that provides *processed* (second-hand) information — that is, information gained from other people's experience or observations, for example, news reports, books, television programs like 'Beyond 2000' and so on.

PRIMARY SOURCES	SECONDARY SOURCES
Personal experience	News reports
Observations	Magazines
Surveys (questionnaires)	Books and encyclopedias
Photographs	Product brochures
Interviews	Video tapes
Excursions	Audio tapes
Guest speakers	Newspapers
Other people	TV and radio programs

Fig. 9.3   Magazines, television and radio are three of many resources from which you may obtain information. (Sanyo Australia Pty Ltd)

If you are organising interviews, questionnaires or other information-gathering activities which involve contacting other people, you have to be well organised. You may need to make telephone calls or write letters to businesses, stores or government departments to ask for information or to make and confirm appointments to see people. (There is more about using the telephone and writing letters on pages 62-65.)

GATHERING INFORMATION AND DATA

The resources you use can make a big difference to the value of information you obtain. If you use the first magazine or book that you find, or speak to only one person, you will not gain a broad understanding of the subject of your investigation. You might finish up with information that is incomplete or misleading, and this could affect your production work and/or your final report.

It is advisable to check the information from the resources you find. In many cases, different sources will deal with the same issue in a different way. You will need to sift through the information and evaluate it first to judge whether it is relevant to your needs.

Select your resources carefully. It is not possible to use *all* of the possible available resources. If you have a particular area you feel comfortable with — for example, conducting surveys — then use it.

TAKING NOTES

When gathering information from books and other texts, DO NOT copy large sections. Jot down the key words and write notes on points of interest. At a later date you can rewrite the information using your own words. If you *do* need to quote from a text, always acknowledge that it is a quotation (use quotation marks), and provide full details of the source (see 'Presenting your Report' below).

It is impossible to write down everything when interviewing, listening to a guest speaker, or watching a video or television program, so you need to jot down the key points and words and then go back over these as soon as possible after the session.

RESOURCES

Following is a list of resources that could be used to assist in your investigation work.

- People
- Places
- Objects
- Texts
- Pictures
- Statistics

PEOPLE

People are the most readily available resource you can find. Older people are especially useful and interesting resources as they have experienced many changes in technology in their lifetime. There are many other people who could assist you.

teachers	parents	sister/brother
relatives	friends	neighbours
managers	employees	consumers
business owners	politicians	trade unions

Any of these people can provide useful information. They could be interviewed or invited to take part in a survey.

CONDUCTING INTERVIEWS

Before you interview anyone, be sure of what you actually want to achieve in the interview. Make a list of relevant, to-the-point questions and try to avoid questions which will produce a simple Yes or No answer. Try the questions out on a classmate or ask your teacher to check them.

Fig. 9.4 When conducting interviews, be sure the person is available and not rushed for time.

If you plan to use a tape recorder, make sure — *before* the interview — that the interviewee has no objections to this. Also, check your equipment before setting out to do the interview. There is nothing worse than having everything ready and finding the batteries are flat! If you intend to make notes (later) from the taped interview, allow plenty of time for this.

If you are not using a tape, be sure to write the answers to your questions *accurately* and neatly so you can read them easily later.

(Before going into a session to hear a guest speaker, prepare a few questions to ask so you get the most out of the occasion.)

CONDUCTING SURVEYS

When conducting a survey, make sure your questions will provide the information you really want. It is important, too, that you survey a broad enough group of people to gain a good overall view of the topic, and to be sure that your conclusions will have real value.

Do not ask 'pointed' questions in an attempt to get the answers to agree with some opinion you already hold. Your survey should be used to gain information, not to prop up your own ideas. Your conclusions and judgements come after you have analysed the data.

A good survey is one which has three or four optional answers for each question. Multiple choice questions allow for a broader range of answers and make it less restrictive on the person being surveyed. For example:

TICK THE APPROPRIATE BOX.

When you have finished with your plastic shopping bags, do you:

☐ throw them in the rubbish bin?

☐ use them for putting other rubbish in?

☐ place them in a recycling bag for collection?

☐ other action (please specify)

....

....

On completion of your survey, you will need to *collate* the information, *analyse* it and *draw conclusions* from the data. For details on how to *present* the data (graphs, tables, etc.), see 'Preparing your Report' below.

PLACES

Places where you can find information include: libraries, school, home, businesses, factories, shops, farms, museums, restaurants, sporting facilities and, of course, the natural environment.

LIBRARIES

The school or local community library is an obvious place to go for information and is a valuable resource if used correctly. The three important steps to remember when using a library are:

1   Locate the information.

2   Make sure it is what you require.

3   Record the information and details of the source.

If you are unsure about how the library system works, ask for assistance from the staff. They will be happy to help. It is silly to waste time hunting around, aimlessly looking for information. After borrowing materials make sure you return them by the due date. Remember, there are many other students requiring the use of them.

OBJECTS

By looking at various items, pieces of equipment or articles, you can gain ideas on how to improve them or how they may be used to assist you in making your own product. They may prompt alternative ideas. They can provide other valuable information about, for example, the type of people who use them, how they affect people's lives, how they are made (whether they are handmade or mass produced), what materials they are made from, and so on.

TEXTS

Books, magazines, journals, newspapers, advertising leaflets, promotional materials, manufacturers' catalogues and other forms of writing are all valuable tools for learning and gaining information. Apart from word of mouth, writing is the most common method of communicating ideas and passing on information.

Fig. 9.5   Use books, magazines and other texts to find information.

Be sure to use resources that are easy to read and understand. Some books and magazines are written in very technical language but technical books sometimes have a *glossary* — a list of special words and their meanings. Check these and use dictionaries, or ask your teacher or other people to explain difficult words.

PICTURES

That old saying, 'Seeing is believing' makes a lot of sense in technology education since much of the work is of a practical nature and involves images and objects. You will often find it easier to understand information if it is represented pictorially. There are many visual resources which provide information. Look for relevant videos, photographs, drawings, graphic representations (charts, graphs, etc.), films and TV programs.

STATISTICS

Statistics are the numerical data (the figures) related to a specific subject. In tennis matches, statistics compiled during the game can tell a broadcaster, for example, how many first serves a player has made or how many aces have been served. These statistics assist in showing up the weaknesses or strengths of the players as well as being informative for the viewer or listener during the game.

Statistics might be used in an investigation for finding out which of the latest CDs had the largest sales during a particular month. This could be achieved by ringing local sales outlets and asking for records of their sales of CDs during that month. Once completed, this would indicate the most popular CD, as well as show which outlet sold the most in the area.

Statistics can be used to assist you with your project work, too. You can find out about people's preferences for materials, or the shape and colour of designs, and so on. Statistics compiled during tests indicate weaknesses and strengths in various aspects of materials or design.

Statistics are compiled from surveys (see above), from records of research data and from various types of tests. After they are analysed, there are a number of ways to represent them. Some examples can be found on pages 58-61.

ANALYSIS, EVALUATION AND CONCLUSIONS

Once you have gathered all your information, you need to sift through it in order to prepare your report. You need to analyse and evaluate each piece of information and data to see how it fits into the requirements of your investigation.

At this point you may discard some material because it is irrelevant, or you may have found more up-to-date information since you recorded it. (But make sure you include it in your report's list of resources, to show the extent of your research.)

After evaluating all the material, you will be in a position to draw some conclusions about the subject. You can then work out how best to present the material to show how you came to those conclusions.

PREPARING YOUR REPORT

There are a number of ways that you can present the findings of your research. It could be in one or more of the forms listed below:

- Written report
- Video report
- Class demonstration
- Written and pictorial report
- Oral report
- Audio tape report
- Poster
- Role play

The length or extent of the report may depend on the requirements of your particular course or teacher, or your own needs. Make sure you know of any special requirements before you start your investigation. No matter which form you decide to use, you must present the report in a well-organised manner.

WRITTEN REPORT

A report consisting of text *only* can often be boring for the reader so you will have to make it interesting. Stick to the point and don't babble on. This will keep the reader (your teacher) keen to read on — and more likely to give you a good assessment. With this type of presentation it is doubly important to make it visually attractive, with neat borders, headings which stand out, and even highlighting to draw attention to the main points where possible. (See also 'Written and Pictorial Report'.)

ROLE PLAY

Role playing is an interesting and enjoyable way of presenting your investigation findings. For example, you could do a mock interview, with yourself as a TV personality interviewing a classmate who is playing

the role of a person from industry or a government department. The interview could deal with a number of issues, for example, health and safety in the workplace. There are many other roles you could act out as well; your imagination is all that will limit you.

VIDEO REPORT

When using video to present your report be sure you have learnt to operate the equipment first. And before you start, make sure the equipment is working. It would be a shame if you put the effort into organising the video to find it did not work properly. If you are using the video for an interview, make sure the topic is interesting and remember to speak up — don't mumble — and try to act naturally. Also, make sure you have permission from the interviewee (the person you wish to interview) *before* you start filming.

ORAL REPORT

If you choose to give an oral report, you will need to be sure of your information so you will be confident enough to stand in front of people to deliver it. Supportive material such as photographs or demonstrations can also assist in this type of presentation.

CLASS DEMONSTRATION

A well-prepared demonstration can achieve a great impact on people, especially if it is a functional demonstration such as showing how an infra-red ray works in an alarm system, or the use of the sun as a source of energy.

POSTER

The poster is a very good format for presenting reports because it is visual and most people tend to learn quickly from visual displays. The use of photographs, drawings and sketches, and a neat layout with careful use of colour make the poster format an acceptable and effective option for getting your message across.

AUDIO TAPE REPORT

The audio tape could be used to assist you when doing an oral report or it may be used as the major presentation tool. It would be wise, however, to incorporate other media as well. Listening to an audio tape can become very boring. The audio tape can also be very useful if you wish to present the views of another person, gained through a recorded interview.

WRITTEN AND PICTORIAL REPORT

The combination of text and pictures is probably the most widely used option. When presenting your report you can put together written information with a variety of graphic representations and other pictorial material: diagrams, photographs, charts, tables, sketches and various other illustrations. Good layout and neatness in this type of presentation are imperative if you are to finish with a quality report.

REPRESENTING STATISTICS

Statistics can be represented in many ways which make the information eye-catching and easy to understand. Some are listed on page 58. (During your research, keep your eyes open for others.)

- bar graph
- line graph
- pie chart
- bubble chart
- tables
- flow chart

The examples below show the use of a few of these to represent data gathered from surveys.

Example 1

Michael was a student of technology at Silver Gully Secondary School. He was required to research and report on problems associated with litter in the local community. Michael used a map, tables and pie charts to show the distribution of the litter, and the relative amounts of the various types and forms of materials found. (See Figs 9.6 – 9.9.)

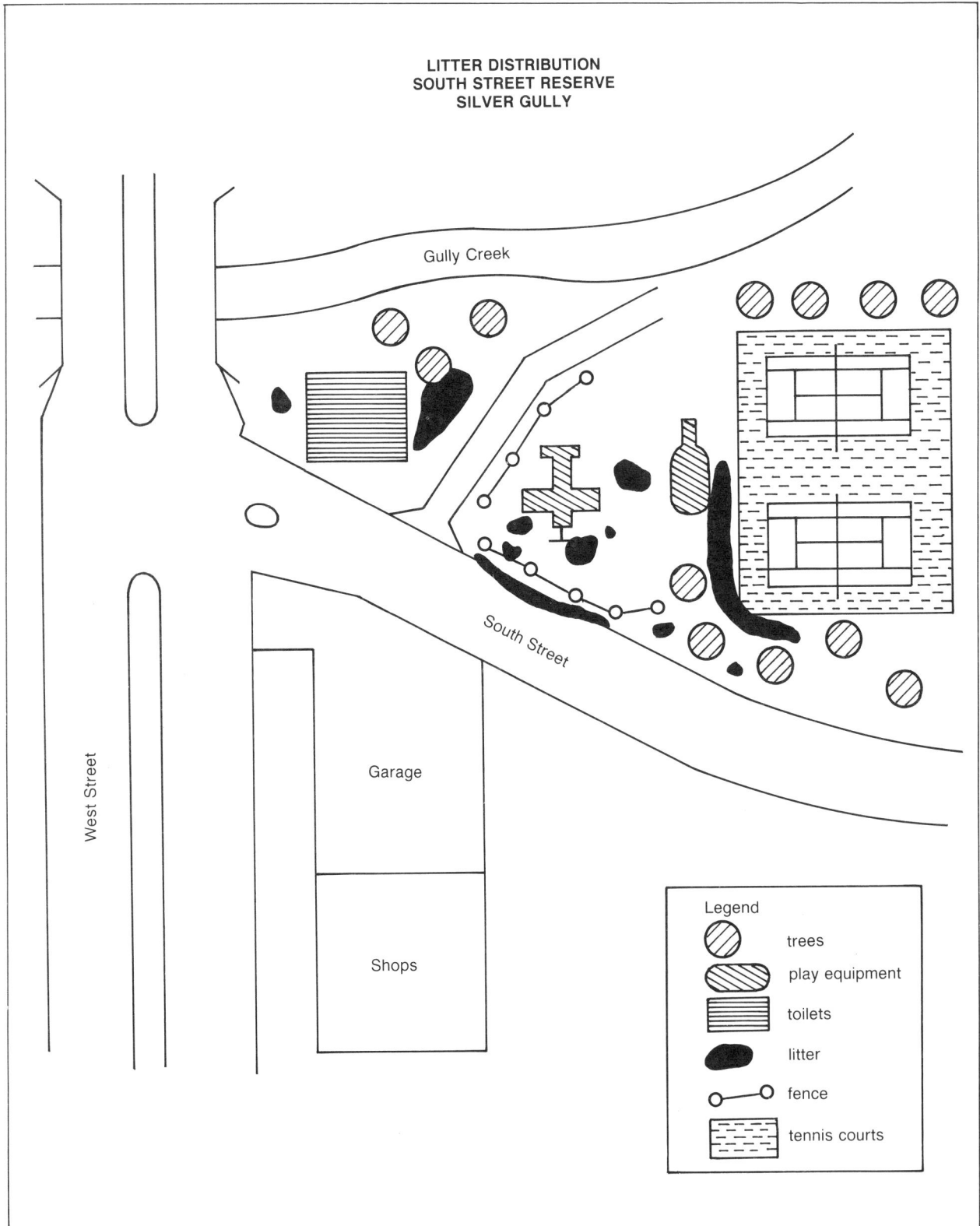

Fig. 9.6

LITTER SURVEY TABLE

	DESCRIPTION	ITEMS	
		Raw No.	% of total
MATERIAL	Paper	130	33.9
	Plastic	147	38.4
	Steel	1	0.3
	Aluminium	57	14.9
	Glass	32	8.4
	Wood	4	1.0
	Cardboard	12	3.1
	Fabric	1	0.3
		Raw total =384	% = (Raw no. x 100) / (Raw total)
FORM	Bags 	42	10.9
	Wrappers	166	43.2
	Newspapers	4	1.0
	Bottles	63	16.4
	Straws	18	4.7
	Cans	11	2.9
	Bottle Tops	35	9.1
	Icy Pole Sticks	4	1.0
	Packets	7	1.8
	Cloth	1	0.3
	Box Containers	32	8.3
	12-Gallon Drum	1	0.3
		Raw total =384	% = (Raw no. x 100) / (Raw total)

Fig. 9.7 The raw figures and percentages are shown on this table.
The percentages are shown again in the form of pie charts.

LITTER SURVEY
SILVER GULLY
MATERIALS FOUND

Paper 33·9%
130

Cardboard 3·1%
12

Wood 1·0%
4

Plastic   38·4%
147

Glass 8·4 %
32

Aluminium 14·9%
57

Steel   0·3%
1

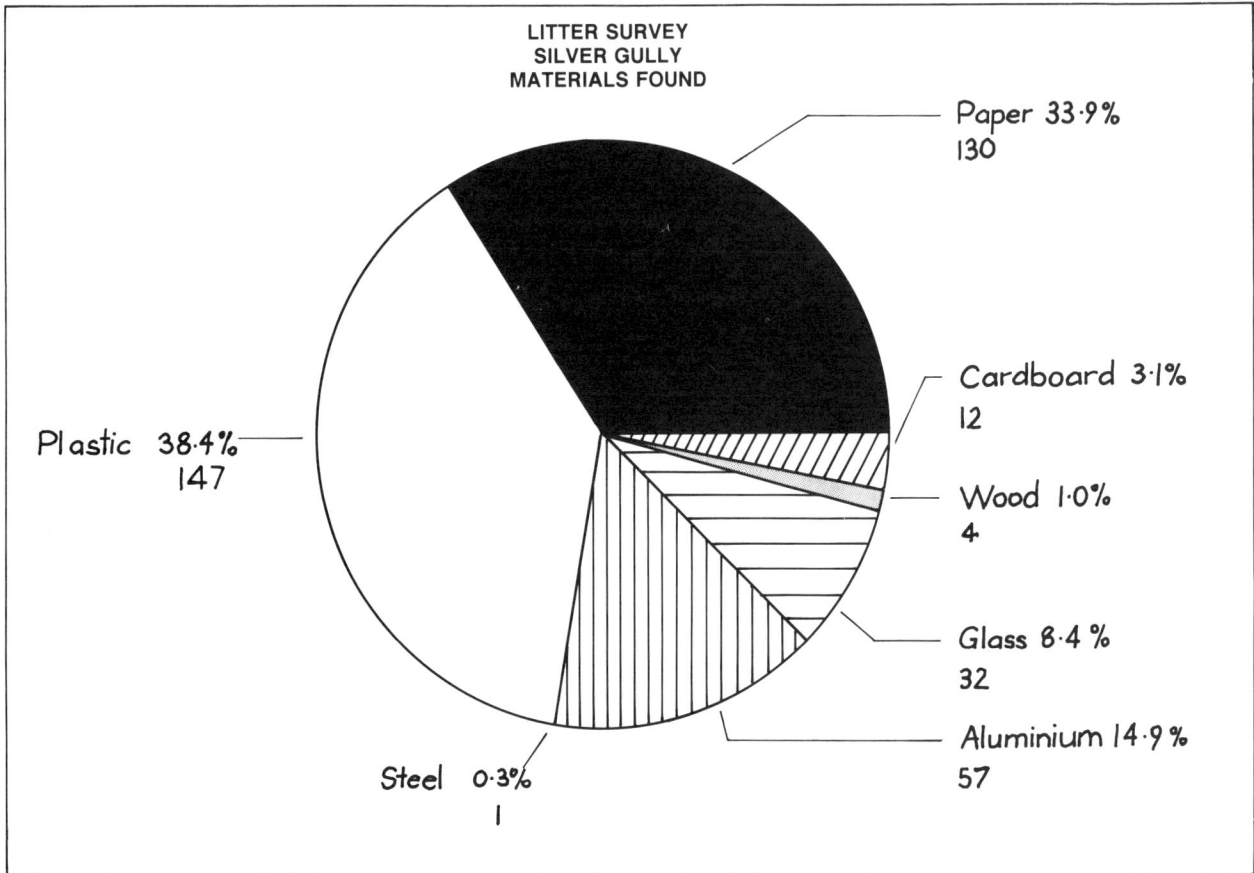

Fig. 9.8   Pie chart showing relative amounts of materials found

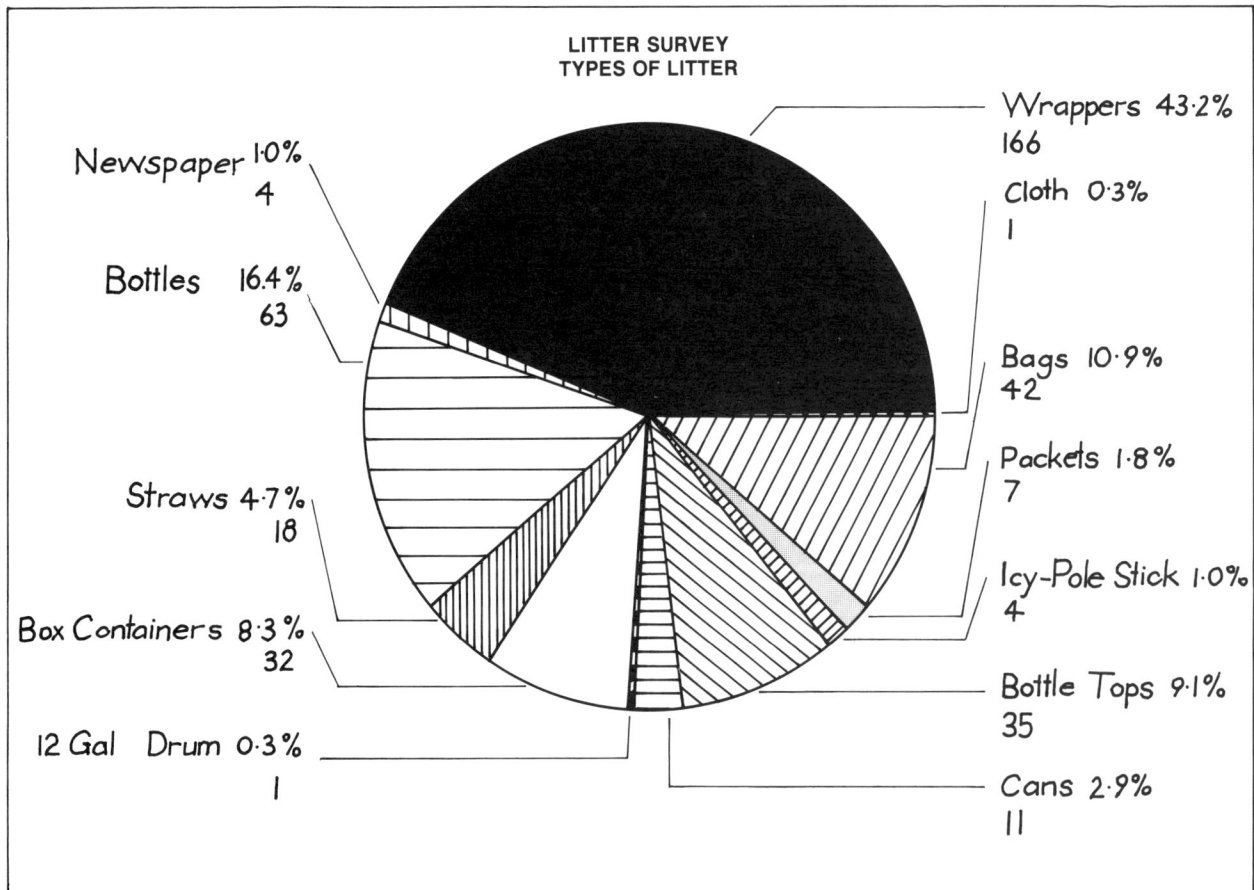

LITTER SURVEY
TYPES OF LITTER

Wrappers 43·2%
166

Newspaper 1·0%
4

Cloth 0·3%
1

Bottles   16.4%
63

Bags 10·9%
42

Packets 1·8%
7

Straws 4·7%
18

Icy-Pole Stick 1·0%
4

Box Containers 8·3%
32

Bottle Tops 9·1%
35

12 Gal Drum 0·3%
1

Cans 2·9%
11

Fig. 9.9   Pie chart showing relative amounts of types of litter

Example 2

After identifying the properties of various plastics, Donna put the information into a table (see Fig. 9.10). This information would be very useful in the selection of the appropriate plastic for a later project. The table format makes the information easy to read at a glance.

Example 3

Ahmet is studying technology in a secondary school in Melbourne. He was required to find out which mode of travel 245 students in Year 11 used to travel to school, and to present the information in a neat and easy-to-read form. Ahmet chose a five-question survey and put his findings together as shown in Fig. 9.11.

Plastics Identification

Name: Gary Dee

Date: 5 May

Characteristics		Polyvinyl chloride (PVC)	Polypropylene (PP)	Polycarbonate (PC)	Polystyrene	Acetal	Low Density Polyethylene (LDPE)	Polymethyl Methacrylate (Acrylic)	High Density Polyethylene (HDPE)	Acryloniterile-Butadiene-Styrene (ABS)	Polyamide (Nylon)	Unplasticised Polyvinyl Chloride (UPVC)	Bottle Base ✳	✳
Touch (Bending)	Rigid			✓	✓	✓		✓		✓		✓		
	Flexible	✓	✓				✓		✓		✓		✓	
(Sound when dropped)	Dull						✓				✓	✓	✓	
	Metallic	✓	✓	✓	✓	✓		✓	✓	✓				
Density	Floats					✓	✓		✓				✓	
	Sinks	✓		✓	✓			✓			✓	✓		
Burn Test														
Flame	Yellow	✓		✓	✓		✓			✓		✓		
	Yellow/Blue		✓					✓	✓		✓		✓	
	Other					✓								
Smoke	Black	✓	✓	✓	✓						✓	✓		
	White										✓			
	Little or none					✓	✓	✓	✓				✓	
Condition	Drips			✓		✓	✓	✓	✓		✓		✓	
	Goes clear		✓				✓	✓	✓					
	Splutters	✓		✓	✓			✓		✓				
	Carbon	✓		✓	✓					✓	✓	✓		
	Self extinguishing	✓		✓								✓		

Fig. 9.10

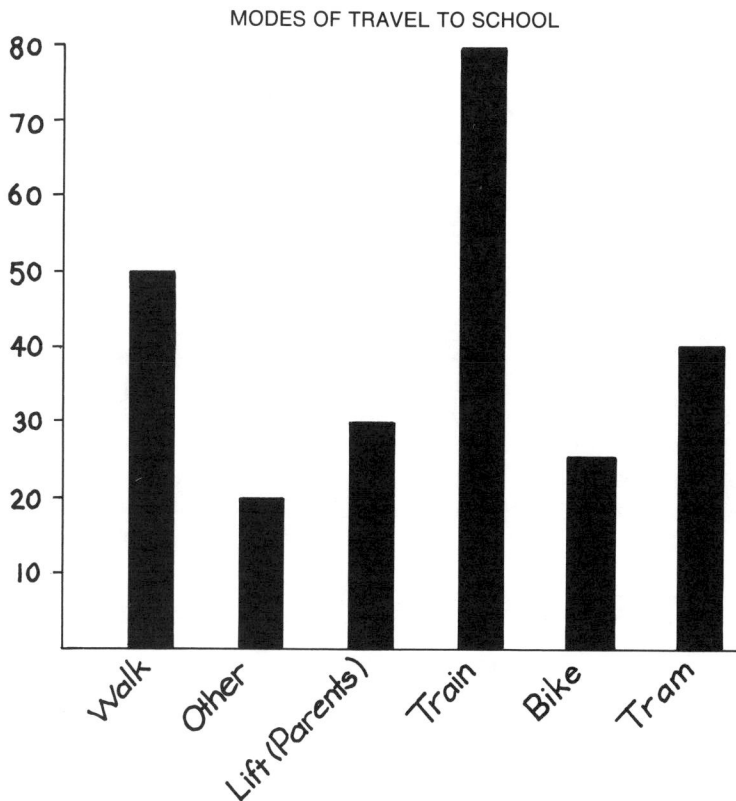

MODES OF TRAVEL TO SCHOOL

Fig. 9.11

PRESENTING YOUR REPORT

WRITTEN REPORT

A suggested format for a written report or a combined written and pictorial report could be as follows. Each heading indicates the *part* you should include and is followed by the *contents* of each part.

TITLE PAGE
Your name
Title of the report
Teacher's name
Date

CONTENTS PAGE
Write down the contents of the report, listing the various sections of the report. Include the page numbers, too.

INTRODUCTION
Write a brief statement about what is being researched and why. Make it as interesting as you can so the reader will be keen to read about your findings.

TEXT (the main part)
Break the topic into sections. Introduce material in a logical way. Support your views with evidence. Use a variety of media (photographs, drawings, charts, etc.).

CONCLUSION
Present your findings. Write something like:
I found that ....
Write a statement on the findings (a paragraph is plenty):
In conclusion, I feel that ....

REFERENCES AND RESOURCES
List all the sources of your information, including details of books consulted, people you interviewed, and so on. For example:

> Smith, K.P. Materials and Design, Daily Publishing Company,
>   Australia, 1990 (pages 56-78).
> Recycling Plastics (on 'Beyond 2000'), Channel 7,
>   27 May 1990.
> Ms Betty Marlow, Managing Director of Plasticworld,
>   Hilltown (interview 4 July 1991).

APPENDIX OF DATA
Attach all support material you may have used in compiling your report: question sheets, letters, newspaper cuttings, statistics and so on.

OTHER REPORTS

For all other forms of report (oral, video, poster, etc.) you should prepare a *summary sheet* to hand to your teacher when you give your presentation. Your teacher can make notes on this and file it. It should include the following:

- your name
- the title of the report
- teacher's name
- date
- a statement saying what form your presentation took
- a list of references and resources
- an appendix of data.

USING THE TELEPHONE

The manner in which you use the telephone when ringing people to obtain information or to make an appointment should always be polite and to the point. Identify yourself straight away and state clearly what you want.

A phone call to a company to gain information on the materials used in its products might go something like this:

TELEPHONIST:	Brunswick Plastics Incorporated. May I help you?
STUDENT:	Hello, my name is Holly Pappa. I am a Year 11 student doing Technology Studies at Oakville Secondary School and I need some information about a product that your company makes. Could you help me with this please?
TELEPHONIST:	Yes, Holly, we should be able to assist you. What exactly do you wish to know?
STUDENT:	I'd like to find out the type of plastic you use to make shampoo bottles, and also how it is manufactured.
TELEPHONIST:	I will send you our product pamphlet. It will give you all of the information you require. May I have your address, please?

In a clear voice, Holly gives the switchboard operator the address and then finishes the phone call:

STUDENT:	Thank you very much for your help. I'm very grateful. Goodbye.

WRITING LETTERS

On many occasions, you will need to write to various companies, factories or businesses to obtain information, or to arrange or confirm an appointment with a member of staff.

Always be clear and precise about what you are wanting. The person reading your letter doesn't need to be given the inside story about your course of study. They need to grasp — quickly — the nature of your request. It is no use writing to a large company such as BHP and asking for 'some information on mining': you must be more specific. If you are looking for information related to BHP's position on mining and the environment, then ask for that. If you can specify a *particular* environmental issue, so much the better.

If possible, use a word processor or typewriter to write your letter. It creates a better image and makes it easier for the person reading the letter to understand what is required.

Of course, after your dealings have been completed, a thank-you letter must be sent to the person you contacted. It is not only courteous and normal business practice: it reflects well on you and your school, and will ensure future cooperation for other students.

SAMPLE LETTERS

Figures 9.12, 9.13 and 9.14 are examples of three kinds of letters: a request for information, a letter of confirmation, and a thank-you letter.

In Fig. 9.12, Brooke is writing to Pylex Plastics Industries in New South Wales. She is doing research on the use of plastics in the garden and needs some information about the type of plastic used for garden hoses. She also wants to know why some hoses crack and break more easily than others.

John Watts has already spoken by phone with Mrs Winter of Tintoys Incorporated, and made tentative arrangements to visit the factory. He writes to confirm the arrangements (see Fig. 9.13). After the visit, he writes to thank Mrs Winter (see Fig. 9.14).

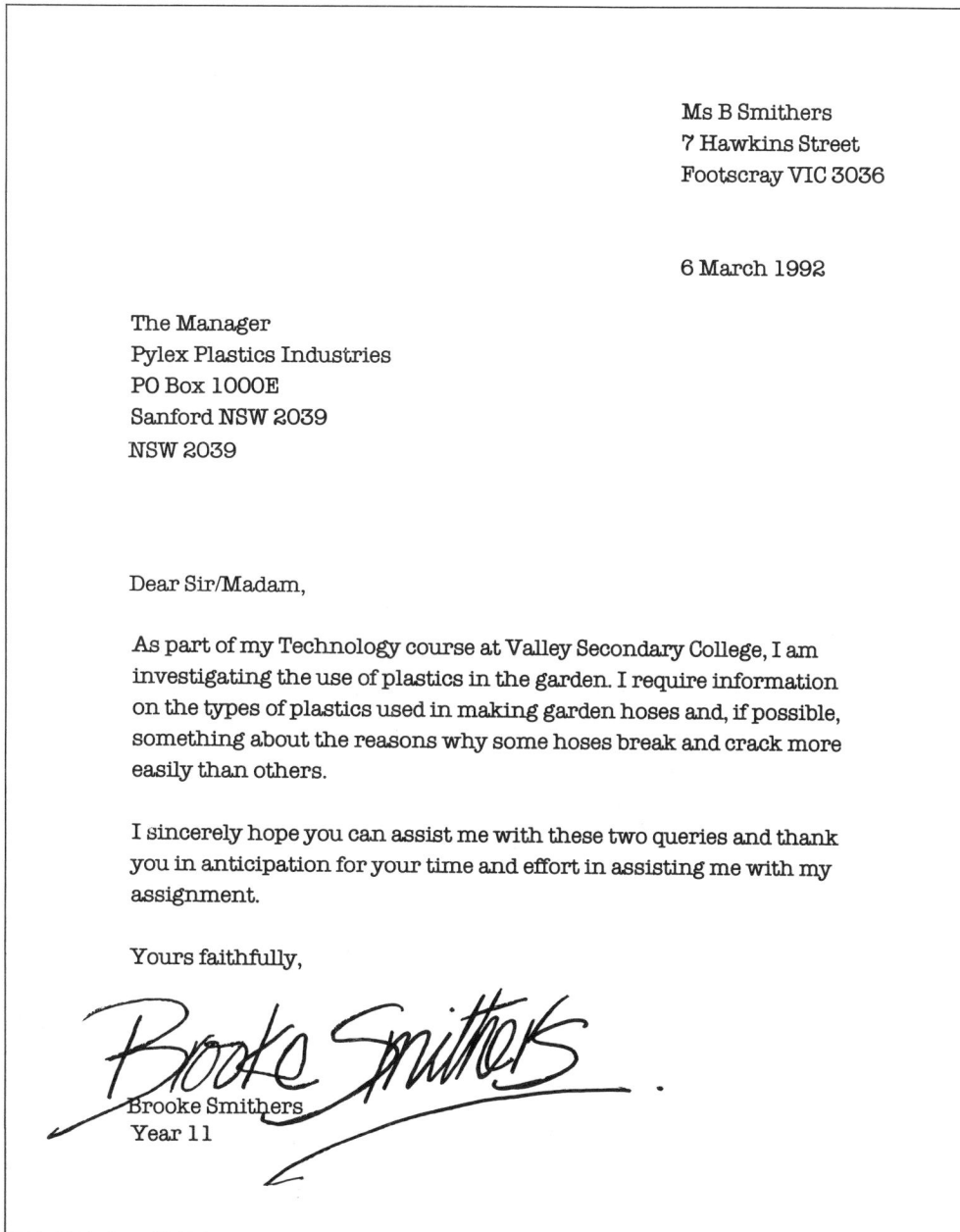

Ms B Smithers
7 Hawkins Street
Footscray VIC 3036

6 March 1992

The Manager
Pylex Plastics Industries
PO Box 1000E
Sanford NSW 2039
NSW 2039

Dear Sir/Madam,

As part of my Technology course at Valley Secondary College, I am investigating the use of plastics in the garden. I require information on the types of plastics used in making garden hoses and, if possible, something about the reasons why some hoses break and crack more easily than others.

I sincerely hope you can assist me with these two queries and thank you in anticipation for your time and effort in assisting me with my assignment.

Yours faithfully,

Brooke Smithers

Brooke Smithers
Year 11

Fig. 9.12

Brunswick Secondary College
114 Long Avenue
Brunswick VIC 3050

26 February 1992

Mrs R. Winter
Tintoys Incorporated
Cold Street
Brunswick VIC 3050

Dear Mrs Winter

I refer to our telephone conversation of Thursday 20 February.
The arrangements we made then have been approved by my teacher,
Ms Brown, and so I will be pleased to visit your factory at 1 p.m. on
Wednesday 11 March as planned.

In our phone conversation I mentioned the use of a tape recorder
for interviewing purposes. At the time you had no objections to this.
If your position on this matter has altered could you please inform
me.

I am looking forward to visiting your factory and gaining much
valuable information about the production of children's toys.

Yours sincerely

John Watts
Year 11

Fig. 9.13

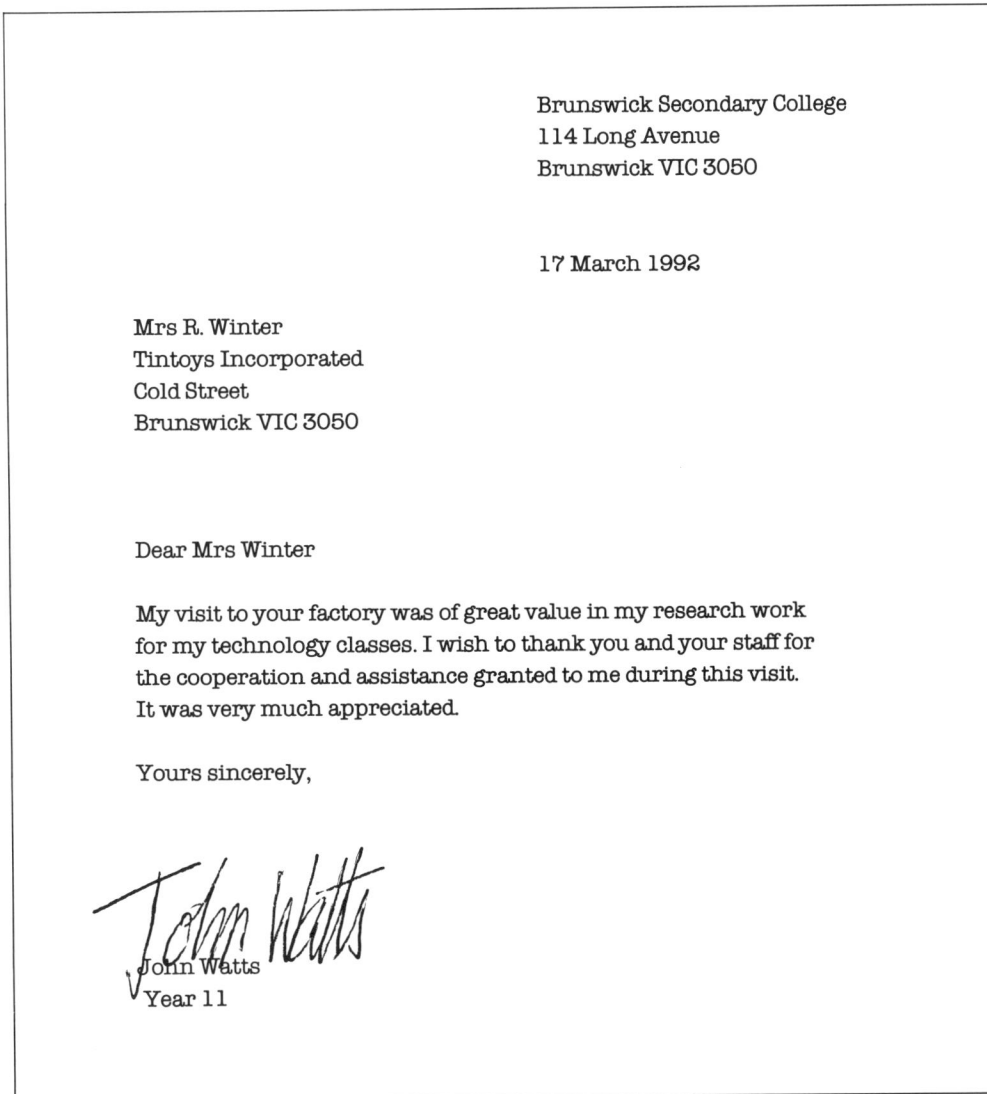

Brunswick Secondary College
114 Long Avenue
Brunswick VIC 3050

17 March 1992

Mrs R. Winter
Tintoys Incorporated
Cold Street
Brunswick VIC 3050

Dear Mrs Winter

My visit to your factory was of great value in my research work
for my technology classes. I wish to thank you and your staff for
the cooperation and assistance granted to me during this visit.
It was very much appreciated.

Yours sincerely,

John Watts
John Watts
Year 11

Fig. 9.14

REVISION QUESTIONS

1 Name the *two* main purposes of carrying out investigation work.

2 List *six* different ways in which you can present your reports.

3 Why is it important to plan ahead and be organised when doing investigations and reports?

4 Name *four* resources which you can use to obtain information for your assignment or report.

5 Name *two* primary resources and *two* secondary resources.

6 Name *three* ways in which statistics can be graphically represented in reports, and explain why they are used.

7 Circle TRUE or FALSE:
   a Investigations help in the making of products.   T   F
   b The library is used only when you require information from books.   T   F
   c The radio is a primary resource.   T   F
   d Letters of thanks are sent to firms only if they ask for them.   T   F

8 Why is it important to write down key points when listening to speakers or watching videos and demonstrations?

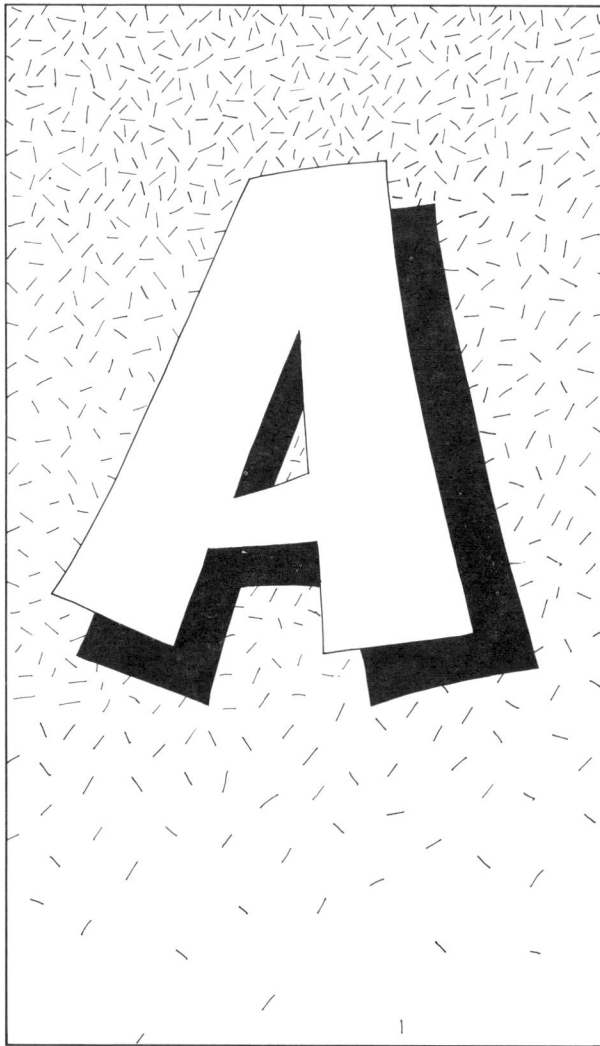

APPENDIXES

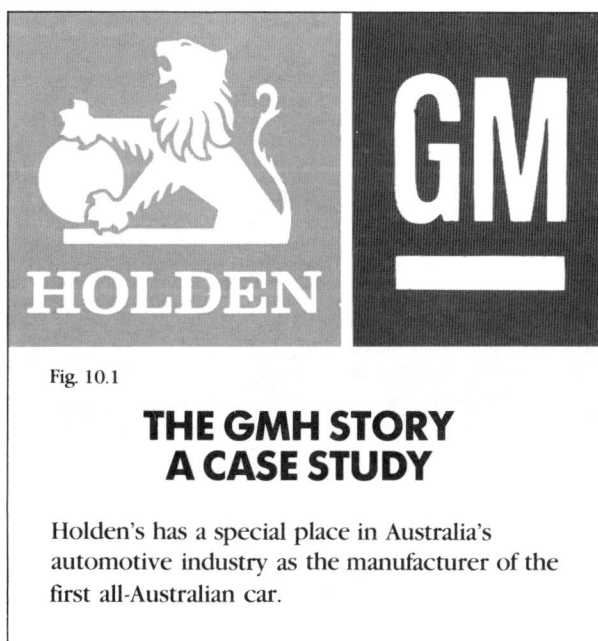

Fig. 10.1

THE GMH STORY
A CASE STUDY

Holden's has a special place in Australia's automotive industry as the manufacturer of the first all-Australian car.

The first Australian Holden car came off the production line in November 1948. The story of the Holden did not, however, begin there. In the mid-1800s, James A. Holden opened a leather and saddlery shop in Adelaide. By 1885 the 'Holden and Frost' company had been formed. Its activities included the making of carriages and coaches but the coming of the automobile soon brought changes.

From the early 1900s, Holden and Frost were assembling automobiles from imported parts. In 1917, the company manufactured its first motor body and soon became the sole supplier of car bodies for the big American company, General Motors Corporation. In 1920 Holden and Frost became Holden Motor Body Builders and, in 1931, this company merged with General Motors to become General Motors-Holden Limited, thus forming the first automotive manufacturing facility in Australia. Many Australian companies owe their development to the initiative and enterprise displayed by Holden's in those early years.

The first discussions to plan the production of an all-Australian car began in 1936 but the plans had to be shelved when the Second World War put a temporary hold on all car production and, like many others, the GMH factory was used for the war effort. War-time production included the making of aero, marine and torpedo engines as well as tools, boats, pontoons and troop carriers. The end of the war allowed the recommencement of work on the local car project (called Project 2000).

Engineers travelled to America in 1946 to study the designs and manufacturing processes used in GM's plants, and returned with three Holden prototypes. The first full production Holden, the 48-215 model, left the assembly line on 28 November 1948.

Fig. 10.2   The first full production Holden — Australia's 'family car'

Since that first Holden passed through the doors of GMH, over 5 000 000 cars have been produced under the Holden name. These include the popular versions of the FJ (1953-56), the EH (1963) and the HQ (1971) — the largest production run of all — through to the Caprice of the 1990s. Now, as part of the Federal Government's 'car plan' to make our automotive industry more efficient and internationally competitive, GMH and Toyota share model production.

Variations in design, and changing materials and manufacturing processes, have played no small part in the story of the Holden. Not only has the shape of the body changed dramatically; so, too, have driver comfort and safety, the mechanisms and electrical systems that power the car, and numerous other components and accessories. Note the changes in the models shown in the photographs.

Fig. 10.3   FJ Holden (1950s)

Fig. 10.4   EH Holden (1960s)

Fig. 10.5   Kingswood (1970s)

Fig. 10.6   Commodore (1980s)

Fig. 10.7   Caprice (1990s)

HOLDEN AND THE DESIGN PROCESS

For a tool cabinet, track suit, bedroom storage container or one of the latest model Holden cars, the actual processes carried out in designing, testing and making the product are very similar.

In many cases, the finished product we buy from the local Holden dealer is very different from the one that was initially designed. Each newly-designed car undergoes many design changes and stringent testing before being passed as suitable to be manufactured.

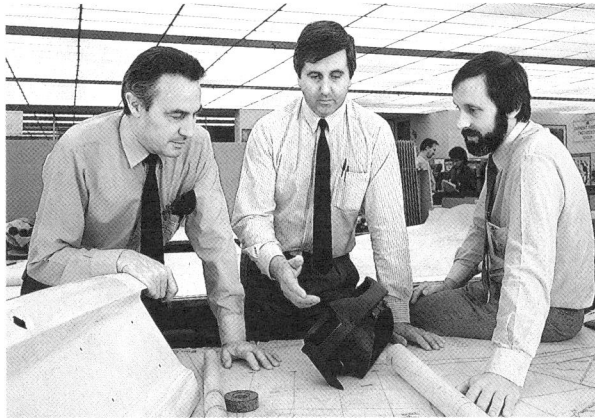

Fig. 10.8   Aesthetics

The steps undertaken to produce the Holden car are not very different from those used to produce thousands of other products. The development of a Holden involves the following steps:

1   Initial ideas and designs are produced — for the inside as well as the exterior — by industrial designers.

Fig. 10.9   a   Early design ideas

Fig. 10.9   b   CAD design

2   Various designs for the body and internal parts are investigated and all aspects researched: material selection, ergonomics, aesthetics, cost, and marketing appeal. Solutions are then developed either through drawing plans and/or making up prototypes (functional models).

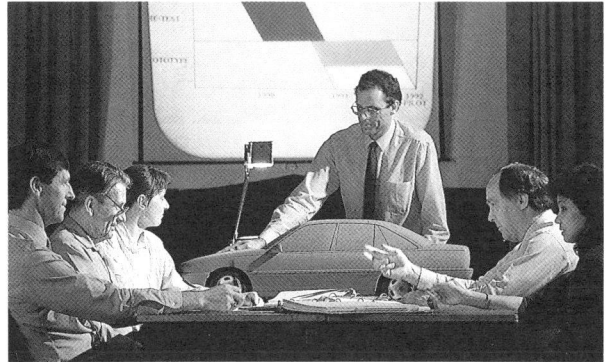

Fig. 10.10   Prototype

3   The functioning of parts is tested. This includes testing for safety, comfort, road handling, cooling, and all electrical and mechanical functions of the car.

Fig. 10.11   Ergonomics

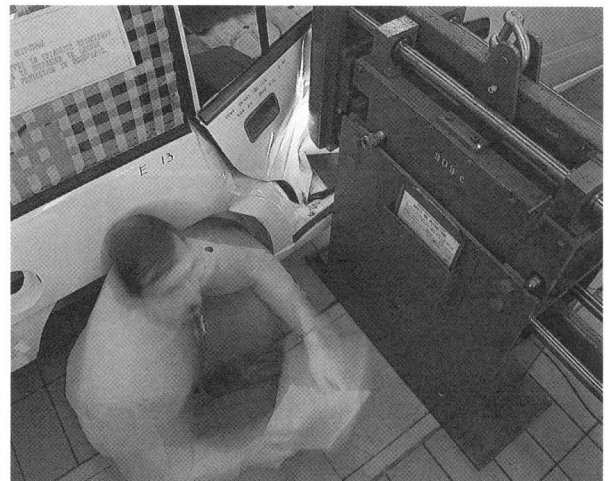

Fig. 10.12   Materials testing

Fig. 10.13 Production

4   The product is manufactured using a wide range of tools, machinery and assembly processes.

5   Product evaluation is carried out to ensure the car functions to the specifications and criteria as laid out in the initial brief.

As you can see, the process is basically the same as the one you have followed in your own work:

● Develop and investigate ideas.

● Design and draw ideas.

● Test and assess the solutions and ideas.

● Make the product.

● Evaluate the product.

STUDENT ACTIVITY

In your school or local library you will be able to find a number of books which have information on the various forms of transport used by people today. List four different types of transport and trace their origins, including information about the people who were involved in their development.

WORKSHEETS

PRODUCT RESEARCH

PRODUCT ....

Company or person responsible for development:

....

When was the product developed? ....

Why was the product developed?

....

....

....

Was it developed especially for Australian conditions? If so, explain.

....

....

....

....

How was the product developed?

....

....

....

Who used (or uses) the product?

....

....

What are the product's strengths?

....

....

What are the product's weaknesses?

....

....

CONCLUSION

....

....

....

....

....

JOURNAL/LOGBOOK ENTRY

NAME .... DATE ....

PROJECT..

Work done this week/session:

....

....

....

....

....

....

....

....

....

....

Alterations:

....

....

....

....

....

....

....

Plans for next week/session:

....

....

....

....

....

....

....

....

....

....

Teacher's signature: ....

DESIGN BRIEF

PROBLEM OR NEED ....

....

....

CONSTRAINTS (LIMITATIONS)

Materials: ....

....

....

Cost: ....

Size: ....

....

Tools and Equipment: ....

....

Time: ....

....

FUNCTION AND SPECIFICATIONS

What is the function of the product? ....

....

Who will use the product? ....

....

Where will it be used? ....

....

....

How often will it be used? ....

....

TOOLS AND EQUIPMENT

What tools do I need?..

....

Tools available	Tools not available	Action taken

OTHER THOUGHTS OR NEEDS ....

....

....

DRAWING SHEET

Brief

Material

Properties

Processes

Part name	Number	Material	Length (mm)	Width (mm)	Thickness (mm)

Name

Project

DRAWING SHEET

REMEMBER

- Draw 3 dimensional drawings.
- Put in all dimensions.
- Draw thickness of materials where necessary.
- Write out notes on extra information and details.

Tools and equipment

Brief

Part name	Number	Material	Length (mm)	Width (mm)	Thickness (mm)

Name

Project

DRAWING SHEET

DEPTH

VERTICAL

45°

HORIZONTAL

LINE CHOICE

Material	Size Quantity	Unit cost		Part name	Number	Material	Length (mm)	Width (mm)	Thickness (mm)
		$	c						
	Total								

Tools and equipment required

Project ——————

Name ——————

DRAWING SHEET

Part name	Number	Material	Length (mm)	Width (mm)	Thickness (mm)

Project ————

Name ————

DRAWING SHEET

Drawing hint

Vertical

Horizontal

45°
Depth

Tools and equipment required

Material	Size/Quantity	Unit cost	
		$	c
		Total	

Part name	Number	Material	Length (mm)	Width (mm)	Thickness (mm)

Name

Project

PRODUCT ANALYSIS

NAME .... DATE ....

PRODUCT ....

What is the function of this product? ....

....

Does the product carry out the intended function? ....

Give reasons for your answer.

....

....

....

....

What material(s) is the product made from?

....

What other material(s) have you seen this product made from?

....

Which product is the most costly? ....

In what way(s) does this product affect the environment?

....

....

Did safety features have to be considered in the design of this product?
If so, explain.

....

....

....

If you had to improve this product, what would you do?

....

....

....

Name *two* features of the product which would help to promote or sell it.

1 ....

2 ....

3 ....

I would advise someone to buy/not buy this product because

....

....

....

....

MATERIAL TEST

NAME .... DATE...

MATERIAL TESTED: ....

Purpose of test:

....

....

....

....

....

....

Description of test:

....

....

....

....

....

....

Tools and equipment:

1.. 2..

3.. 4..

5.. 6..

7.. 8..

Test results:

....

....

....

....

....

CONCLUSIONS

....

....

....

....

....

....

....

PRODUCTION PLANNING
(broad plan)

PLAN

1 ....

....

2 ....

....

3 ....

....

4 ....

....

5 ....

....

6 ....

....

7 ....

....

8 ....

....

9 ....

....

10 ....

....

CHANGES/ALTERATIONS   Teacher's
                                                                                                            initials

Date:.................. Change: ....

....

Date:.................. Change: ....

....

Date:.................. Change: ....

....

Date:.................. Change: ....

....

Date:.................. Change: ....

....

Date:.................. Change: ....

....

PRODUCT EVALUATION

NAME .... DATE ....

PRODUCT ....

What were your initial intentions for the product?

....

....

Has the product fulfilled your intentions? If not, how and why did it fail?

....

....

....

....

....

List the alterations you made in the design and production stages. State briefly why these alterations were necessary
(approx. 20 words for each).

....

....

....

....

....

....

....

If you had to make the product again, would you change your approach, the materials, or other aspects?
Explain briefly (approx. 50 words).

....

....

....

....

....

....

Name *two* tools which were essential in the production of your product. Explain their functions and their value to your work.

1 ....

....

....

2 ....

....

....

MATERIALS COSTING

NAME ....

PRODUCT ....

| Part | Material | Quantity | Dimensions | | | Unit | Unit cost $ c | Total cost $ c |
			L	W	Th			
							TOTAL COST: $	

INFORMATION RECORD

DATE..................................

TOPIC ....

....

Source:

....

....

....

Chapter/Volume: ....   Pages: ....
(if applicable)

         ....

         ....

         ....

Type of information:

....

....

....

....

Alloy a new substance formed when two or more metals or elements are mixed together. (Brass is an alloy of copper and zinc. NOTE: *Aluminium,* the silver-white element, is *not an alloy* but it is a component in many alloys.)

Alteration a change or modification to an initial idea or thing

Analyse to examine critically; to determine the constituent parts of something, to examine these

Analysis the process of analysing; an outline or summary

Appearance the look or outward show of something

Appropriate suited to the purpose

Aspects the ways a thing may be looked at or viewed

Assessment judgement, value; (in school) marking or grading

Assignment a particular task or duty

Authentic real; of genuine origin, not a copy or replica

Brief (in design work) the statement of a problem or need and the written notes and instructions from which a solution to that problem or need can be developed

Capability the fitness or ability to carry out a task

Capital the wealth (money, property) employed in the making of more money (Capital + labour = enterprise or business)

Challenge a demand on one's abilities or skills

Collate to compare texts, statements, etc. in order to find agreements and disagreements

Communication the imparting or exchange of thoughts and ideas

Compatible able to exist together in harmony

Complex complicated, not simple

Complicated made up of many parts

Component one part of something made up of many parts

Composite made up of a variety of parts or elements; any material like this

Compress to press closely together; to force into a smaller space

Concept thought, idea or notion

Conclusion the end or final part; a decision or opinion

Constraint a limitation or restriction

Construct to form by putting parts together; to build

Consultation discussion, seeking advice or guidance in making plans

Consumer one who buys goods (commodities) or services

Creative resulting from original thought; having a talent for imaginative ideas

Criteria standards on which judgements can be based (singular = *criterion*)

Cutting lists lists of materials with accurate details of sizes (length, width, thickness) to which the materials are to be cut

Data detailed information; facts and figures on a subject

Deformation the changing of the shape or form of something

Depict to show

Design to outline (with sketches, technical drawings or plans) the ideas and details of a product to be made or manufactured; the plan or outline itself

Designer a person who plans or fashions products or artifacts through artistic or technical sketches and drawings

Detailed showing or giving the particulars

Devising making, planning

Diagram a drawing or plan that outlines and explains the parts or operations of something

Diligent careful and persistent in one's work

Dimension measurement in one direction (e.g. length)

Discover find, come upon

Draft a preliminary form of any written text

Ductile 1 capable of being drawn out into threads (e.g. gold); 2 capable of being hammered out thin into sheet form (e.g. metal); 3 capable of being moulded or shaped (e.g. plastic); 4 able to withstand pressure without fracturing

Efficient operating or performing adequately, particularly without waste of energy or other resources

Elasticity the capacity to recover shape after deformation

Ergonomic (of furniture, etc.) designed for maximum safety, comfort, health and efficiency

Escalating rising, developing, increasing

Estimate to make an approximate calculation; the approximation itself

Evaluate to assess the value of something (an object or situation) by testing it against relevant criteria

Expanding growing larger, extending

Export (v.) send to or sell to another country; (n.) a product which is exported

Fabricate to make, construct or assemble

Fabrication something fabricated

Factors elements that contribute to a given result or state

Fibre a fine, thread-like piece of material

Format the plan or style of something; the way it is presented

Formulate   to state in a precise form
Freehand   drawn by hand, without aids
Function   the special purpose or working use of something

Highlight   make something stand out, bring to the attention
Horizontal   parallel with the horizon, at right angles to the vertical

Identify   to recognise and name
Illustrate   to draw pictures or sketches of an item
Imagination   the ability to create pictures in the mind
Imported   brought in or bought from another country (NOTE: **br**ought goes with **br**ing; b**ought** goes with b**uy**.)
Incentive   something offered as an encouragement to do something
Industry   any production or manufacturing business; any large-scale business activity (e.g. *fashion industry*); manufacture or trade as a whole;
Primary Industry any industry (e.g. mining, farming) which does not manufacture the goods (e.g. iron ore, wool) that it produces; **Secondary Industry** any industry (e.g. steel industry, textile industry) which produces manufactured goods (e.g. tubular steel, cloth) from the products of primary industry
Inferior   not up to acceptable standard
Inhabited   lived in, occupied
Initiate   to begin or originate
Innovate   bring in something new, especially new ways or methods
Innovation   the introduction of new things or methods; the new thing or method itself
Innovative   new, creative
Invent   to make up or think of first; to originate
Investigate   to search or inquire into a subject or topic
Investigation   a systematic search or inquiry into a subject or topic
Isometric   having equal measurements

Journal   a record of findings and writings made during a unit of work (sometimes called a *Logbook*)

Lay person   a non-professional person
Limitation   ` a restriction or constraint
Logbook   *see* **Journal**

Manufacture   to make or produce goods
Marketing   selling; the advertising and public relations conducted to promote products
Measurement   the dimension of something measured

Microchip   complex electronic circuits in a single package
Mock-up   a model of an item or object, usually used to test appearance, etc.
Model   something made to simulate another object
Modify   change or alter

Oblique   slanting; neither perpendicular nor parallel to a given line or surface
Optional   open to choice
Ore   a metal-bearing mineral or rock, especially when valuable enough to be mined
Original   the first of its type, something from which copies are derived

Perspective   a view of an object drawn in such a way as to show dimensions and spatial relations
Pictorial   of or like or illustrated by a picture
Practical   relating to practice or action
Procedure   the manner in which one proceeds through an action or task; the steps involved
Process   a systematic series of actions directed at some end (e.g. the making of a product)
Production   the making of a product
Progressive   proceeding step by step; going forward and onward
Property (of a material)   a distinctive characteristic, quality or ability
Proportion   the comparative relationship of size between things or parts; ratio
Protection (of local goods)   the tariff and import restrictions which keep the price of imported goods high to match local prices
Prototype   the first working model of something

Quality   high-grade or superior on a scale of excellence
Questionnaire   a list of questions used to obtain opinions on some subjects

Redundant   no longer needed or required
Reference   a direction to a source of information
Relevant   to do with the matter in hand
Report   a verbal, visual or written account of the findings of an investigation or observation
Requirement   something required, demanded or obligatory (see *Work requirement*)
Research   to investigate, look into or seek information on a topic
Resource   **1** the source of support, supply, aid or information; **2** the source of potential wealth (e.g. mineral deposits)
Restructuring   the re-organisation of some industries and systems to make them more efficient

Revolution   a complete or radical change; one complete turn of a wheel, etc.
Robotics   the study and technology of robots
Rotary   having a part or parts that rotate

Scale down   to reduce in size according to a fixed ratio or proportion
Scale up   to increase in size according to a fixed ratio or proportion
Selection   a choice from a range of possibilities
Simulate   to imitate (NOTE: Do not confuse with s**t**imulate.)
Skill   the ability to do something well, acquired from knowledge and practice
Smelt   to extract metal from ore by heating, melting, etc.
Specification   a statement of precise requirements
Standard   quality, level of excellence or achievement
Stringent   rigorous (tough) standards of performance
Strive   to try hard

Technique   the particular method or procedure for doing something
Technology   the use of a wide variety of skills, knowledge and creativity in attempting to find solutions for a particular need or problem
Tensile strength   the force needed to stretch a material until it breaks; a material's capacity to be pulled apart
Test   a trial or procedure to determine quality, ability, composition, etc.
Toughness   hardness; the ability not to be easily broken, cut, etc.
Transistor (in electronics)   a miniature solid-state device for amplifying or switching
Tubular   having, consisting of, or shaped like a tube

Unique   having nothing like or equal; the only one

Vertical   perpendicular to the horizon; straight up
Viability   practicability, workability, capacity for success
Visualise   to imagine or form a mental picture of something

Work requirement   any activity or task that is required to be undertaken, often as part of assessment (as in VCE courses in Victoria)